# DEGREES
# OF
# COMPROMISE

SUNY series in

## SCIENCE, TECHNOLOGY, AND SOCIETY

Sal Restivo and Jennifer Croissant
*editors*

# DEGREES
# OF
# COMPROMISE

*Industrial Interests
and Academic Values*

JENNIFER CROISSANT and
SAL RESTIVO, editors

STATE UNIVERSITY OF NEW YORK PRESS

Published by

STATE UNIVERSITY OF NEW YORK PRESS, ALBANY

© 2001 State University of New York

For information, address
State University of New York Press,
90 State Street, Suite 700, Albany, NY 12207

Production, Laurie Searl
Marketing, Dana E. Yanulavich

Library of Congress Cataloging-in-Publication Data

Degrees of compromise : industrial interests and academic values / Jennifer Croissant and Sal Restivo, editors.
         p. cm.
    Includes bibliographical references and index.
    ISBN 0-7914-4901-7 (alk. paper) —
ISBN 0-7914-4902-5 (pbk. : alk. paper)
      1. Industry and education—United States.
    2. Education, Higher—Economic aspects—United States. 3. Universities and colleges—United States.
    4. Research—United States.    I. Croissant, Jennifer, 1965-    II. Restivo, Sal P.

    LC1085.2 .D44 2001
    378.1'03—dc21
                                    00-045678

10    9    8    7    6    5    4    3    2    1

# Contents

# Figures and Tables

## FIGURES

## TABLES

# Acknowledgments

*Science as a Vocation in the 1990s: The Changing Organizational Culture of Academic Science* was previously published in the *Journal of Higher Education* 61(3, May/June 1990):241–279. It is reprinted here with permission. Copyright 1990 by the Ohio State University Press.

*Industry, Academe, and the Values of Undergraduate Engineers* was previously published in *Research in Higher Education* 33(3):275–295, 1992. It is published here with permission.

*Systemic Influences: Some Effects of the World of Commerce on University Science* is reprinted as *Pervasive Influence: Intellectual Property, Industrial History, and University Science* from *Science and Public Policy* 25(2): 95–102, and appears with permission.

*The Gloves Come Off: Shattered Alliances in Science and Technology Studies*, by Langdon Winner, originally appeared in *Social Text* 14(1&2, Spring/Summer 1996):81–91. It is reprinted here with permission.

Cover art from Digital Imagery® copyright 1999 Photo Disc, Inc.

# Introduction

## JENNIFER L. CROISSANT AND SAL RESTIVO

Something is happening to science. We are told of the need for a new social contract for science (Byerly and Pielke 1995). Competitiveness is a powerful, emergent rhetoric that frames research and development policies (Slaughter and Rhoades 1996). Institutions are experimenting with new configurations of government, universities, and business (Bowie 1994). We should perhaps think of science in a new "triple helix" of government-industry-university relations (Etzkowitz 1990) or as moving into a "new production of knowledge" mode (Gibbons et al. 1994). Publication patterns and the location of scientific and technical activities are shifting, while collaborations between disciplines and nations grow (Hicks and Katz 1996). Perhaps we even live in a knowledge society (Stehr 1994), which could have important implications for universities as sites of knowledge production.[1]

However, saying that "something is happening to science" ignored the fact that actors in a wide variety of institutional settings are working to change science and technology research. Similarly, the use of ecological (Byerly and Pielke, 1995 1998) metaphors to describe science funding environments, defuses questions of resource dependency, agenda setting, quality of work experience, and increasing stratification across institutions and disciplines. These questions are made highly relevant by increasingly explicit calls for measurable economic benefits from university activities. Slaughter and Rhoades (1996) traced the legislation that facilitated a research and development policy in the United States based on competitiveness. They documented the involvement of a broad coalition of agencies, including the National Institutes of Health (NIH), the Department of Defense (DOD), Department of Energy (DOE), and the National Aeronautics and Space Agency (NASA), as well as manufacturers associations, contractors, presidential councils, and eventually the National Science Foundation (NSF). From the early 1980s through the present, commercialization of research has been a consensus policy: Not a the natural "evolution" of research and development practices, but a conscious reprioritization by a broad coalition of actors. Jane Smiley (1995) humorously noted this trend in a work of fiction:

> Associations of mutual interest between the university and the corporations were natural, inevitable, and widely accepted. According to the state legislature, they were to be actively pursued. The legislature, in fact, was already counting the "resources" that could be "allocated" elsewhere in state government when corporations began picking up more of the tab for higher education, so success in finding this money would certainly convince them that further experiments in driving the university into the arms of the private sector would be warranted, that actually paying for the university out of state funds was irresponsible, or even immoral, or even criminal (robbing widows and children, etc. to fatten sleek professors who couldn't find real employment, etc.). (22)

The reprioritization of technoscience, in terms of commercialization and competitiveness, emerges in a larger context where "the privileged position of business in policy making" (Lindblom 1980, 71–82) continues as a problem-free orienting assumption for science and technology public policy and for university activities. Debates about public control, private profit, and professional autonomy are clearly not new to research and universities (Noble 1977, 1986; Kevles 1987; Dickson 1988; Leslie 1993; Guston and Keniston 1994, Smith 1990). But administrators and faculty have gradually redefined the public interest as a matter of private-sector economic development (Slaughter and Rhoades 1991), rather than as a matter of traditional autonomy for scientific and technological research activities. Our book title, *Degrees of Compromise*, is selected to highlight the fact that what is occurring in contemporary U.S. science and technology is new only as a matter of degree. The scope and scale, and the transparency with which industrial interests are prevalent in academic institutions, have been transformed by competitiveness agendas, coalitions, and new institutional arrangements. The title also reflects the idea that the institutions of research are also institutions of higher learning. The degrees awarded, their meaning, content, and the prospects for future work and learning, are all important to both the degree seekers and those who confer degrees in these changing times. On the one hand, the shift toward business values in education reflects students' own orientations toward education. This is noted in annual surveys of students reporting increasingly instrumental rather than expressive values for education. They are concerned with skills and wages rather than self-development in the tradition of the liberal arts. On the other hand, given that degrees are possibly compromised by industry-specific demands for labor and training and by business orientations toward efficiency and secrecy, we will find continuing tensions in the relationship between education imagined in the classic sense of self-improvement and citizenship, and education as training for job-specific skills and values. The result will be a compromise between fundamentally unequal partners.

The debates between Senator Harley Kilgore and Vannevar Bush on the structure of postwar research policy in the 1940s and 1950s (Kleinman 1995) illustrate one era in the university-industry-government nexus. The model of autonomy for basic research held fast as narrative and policy for a full generation. In the current environment private interests are conflated with public control, and the necessity for some sort of state involvement is largely a settled question. Some conflicts, nevertheless, continue over the responsibilities of universities to their various constituencies and over questions about the private use of public interests. For example, land grant colleges have been criticized for their apparent capture by agribusiness research agendas at the expense of small farmers and sustainable farming techniques (Busch and Lacy 1993; Lockeretz and Anderson 1993).

Conflicts over intellectual property and secrecy in research have moved beyond the anecdotal and speculative and into popular, scientific, and scholarly publications (Friedly 1996; Ercolano 1994; Rosenberg 1996; Blumenthal et al. 1996). Blumenthal (1996) notes that "the increasing involvement of for-profit biotechnology companies has provided new sources of funding, but this involvement has an emphasis on the ethical and operational rules of business rather than those of science" (393). Similarly, Rosenberg (1996) recounts specific cases from his experiences, such as a stalemate over access to patented reagents. Dosage and toxicity information were to be withheld from researchers unless they promised confidentiality, so that a competing firm could not get the information. Bowie (1994) similarly outlines case studies of conflicts of interest and intellectual property disputes.

In at least a few states, changes in intellectual property statutes and institutional polices have eroded the labor autonomy of the professoriate (Slaughter and Rhoades 1993). Business-derived models of productivity, as opposed to collegial ideals of scholarship, craft, and service, vex researchers and students in increasingly bureaucratized environments. One can imagine the day sometime in the future when a professor will spend more time classifying, measuring, and reporting his or her activities than actually performing them, at least until such time as those administrative activities in themselves become defined as professorial duties. The changes in universities are matters of degree; the compromises are as well.

For the most part these changes have been seen as inevitable and logically and logistically necessary, although welcomed by some and seen as unfortunate by others. As Slaughter (1990) notes, "very close relationships between business and higher education are presented as virtually problem free" (182). New models map the emerging dynamics, generally neglecting critical assessment of either prior forms of science funding or the new configurations. These models "naturalize" the new arrangements with at most a wry irony for lost opportunities, roads not taken by academic institutions, the marginalization of scholarship in the social sciences and humanities, and ethical conflicts that arise (cf. Gibbons et al. 1994; Etzkowitz 1990). Few studies have been done on the potential value changes arising out of the new configurations of university research. Of those, most have reported largely positive results of interactions with industrial and government sponsors, including individual economic gain (cf. Zucker, Darby, and Brewer 1994). Bowie (1994) is a notable exception. It is also clear that standard quantitative indicators of scientific activity (cf. Hicks and Katz 1996)—productivity and economic development—cannot illuminate either the character or scope of these issues (Vavakova 1996). Historical studies have questioned as to whether a university can "cooperate with industry without compromising itself unduly" (Clark 1997, 96). The chapters in this book rely on multiple methods (surveys, fieldwork, extensive interviews, historical research, and qualititative and quantitative analyses) to provide a more complex picture of institutional and values changes in academic life, and the mechanisms by which these changes occur.

Understanding value changes arising out of new university-industry research relationships (UIRR) requires also articulating the values that we see challenged by new policies and practices. Of course traditional values of professional and disciplinary conduct were not without their problems. The norms of science defined by Robert Merton (1973) were not without their internal conflicts and ideological functions (Mulkay 1979) and have not necessarily produced a professoriate with much public accountability. The fact that contemporary U.S. academic institutions are each historically particular embodiments of the "Humboldtian principle" (Clark 1995, 2)—of inquiry-based education that integrates research, teaching, and study—should alert us to some fundamental tensions among the activities and their ethical basis and among those constituencies with a stake in higher education. Given University-Industry Research Relations (UIRRs) and other forms of academic and industry interaction, we expect, and find in the studies that follow, pointed challenges to the norms of communalism (sharing of knowledge), disinterestedness (lack of financial or personal gain from knowledge), and of organized skepticism (knowledge claims interrogated through review by peers and the public).

The transformation of university values from norms centered on professional scientific practice to those based in industrial measures of productivity, utility, or competitiveness suggests that several significant and historically important goals of universities face possible erosion. For example, as noted above, the norms of utility suggest a transformation from traditional ideas about education to more instrumental notions of skill acquisition, job placement, and "useful" knowledge. In addition, ideas about universities as repositories of culture and knowledge, places of speculation or critique, and ideas about academic freedom are challenged and often dismissed in the name of cost effectiveness.

As Slaughter and Rhoades (1996) note in their analysis of a decade of the developing competitiveness agenda for science and technology research efforts, the underlying assumptions of the agenda are shared alike by policy makers, university leaders, and at least some scientists (331). Proponents of new science policies assume that demonstrations of the contributions research and development make to economic competitiveness will lead to increased basic research funding without changing the activities of working scientists, the university at a structural level, or the processes and outcomes of educational activities. In this model, information is supposed to pass more efficiently from sector to sector, and the content and quality of the information is assumed to be of high integrity. This collection is part of the efforts to challenge those assumptions. Several chapters report the results of a National Science Foundation Ethics and Values studies grant to four faculty investigators for an interdisciplinary research project undertaken to understand new forms of university-industry-government interaction and the ethical and value consequences of such new associations. Contributors to the initial NSF proposal include Sal Restivo (sociology/ anthropology), Deborah G. Johnson (philosophy), Edward J. Hackett (sociology), Langdon Winner (political science), and their student collaborators, Jennifer L. Croissant, Juan Lucena, Mark Gaylo, and Blair Schneider.

Jeffrey L. Newcomer, while a graduate student at RPI, focussed on a government agency award—designed to attract industry funds to the research center—and the resulting transformation of research and educational practices. Daniel Lee Kleinman discusses the underlying assumptions about industry in agricultural biotechnology that shape research trajectories. In other contexts, W. Patrick McCray and William Kaghan provide qualitative assessments of technology transfer activities, while Jason Owen-Smith analyzes intellectual property activities and institutional stratification using quantitative tools. We have written this book as a step toward synthesizing these methodological approaches to studying institutional change and ethical issues. Our interdisciplinary approach illustrates a more critical understanding of the consequences of business values and agendas for the values and objectives of university research.

In certain ways, the research reported here is perhaps so specific to particular institutional contexts that our claims for general concerns about accountability, quality of work experiences, learning, and values change may seem out of proportion. For example, Rensselaer (RPI), as a technological institute and a primary research venue for several of the cases, has a unique position in the history of higher education (cf. Noble 1977) and has always been concerned about its industrial connections in one way or another. Engineering, the major research and educational agenda of the institute, has been a profession strongly connected to industry in both theory and practice. Engineering is also a surprisingly weak profession (Collins 1979), given the "industrial exemption" which allows nearly 90 percent of working engineers to forgo licensure. Engineering research has also been very much application oriented and focussed toward commercial interests and entrepreneurship. However, in the spirit of this volume's title, what is new is the unprecedented degree of pressure for soliciting funds from commercial sources, even within a traditionally industry-oriented institution. For a group of faculty already interested in the application of research to practical ends, new ethical dilemmas have been identified (Ercolano 1994). Given that there are unprecedented incursions into academic programs, ethical challenges, conflicts over research agendas, and a loss of autonomy for this institution, which is ostensibly experienced with industrial interactions, we expect that other academic settings will very likely see larger value dislocations and conflicts as the institutions of science, commerce, and the state intersect.

The initial grant from which the RPI studies emerged was based on the then-recent emergence of research centers as a particular form of the university-industry-government nexus. Research centers, as modes of implementation of the larger changes in funding priorities from the state and expectations from industry, both reflect and lead the values changes across the institutions. Two of the three centers studied ethnographically, the Center for Composites, Materials, and Structures (CCMS) and the Space Research Center (SRC), represent government-based mediations of industrial interests. The Defense Advanced Projects Research Agency (DARPA) and the Office of Naval Research (ONR) are at work in the first. In the second, NASA—through the SRC—was trying to interest industry in funding space research by impelling researchers to do work of perceived interest to industry. The goals of these programs were to facilitate industrial use of knowledge produced through agency sponsorship, with the ultimate goal of having the centers be self-supporting on commercial research contracts. In neither case did this occur. Related programs from the NSF (summarized by Bowie 1994, 15–18), which include the Industry-University Cooperative Research Centers Program, have shown similar difficulties in achieving financial independence from the federal government and full support from industry. In the cases studied in this volume, commercial needs for control of intellectual property were a major barrier to complete integration of industrial research with the base of research established by the centers.

The center approach is only one example of UIRR and state interactions. Dooris and Fairweather (1994) describe a range of organized research units (ORUs), including centers, business incubators, consortia, and research institutes that have emerged, almost entirely since 1980. Conflicts over intellectual property, research collaboration, hierarchy, education, and research autonomy are perhaps more acute in a center setting. Dooris and Fairweather argue that increased technology transfer or interdisciplinary ORUs in and of themselves, however do not inevitably lead to value changes. Centers nonetheless represent the range of issues which these tripartite collaborations can manifest.

While the research center is heralded as an institutional innovation in crossing disciplinary boundaries, the conflicts wrought in these border crossings are also manifest in other forms of research sponsorship, such as individual direct contracting, other kinds of centers, and other kinds of collaborations and joint ventures. The contributions from McCray, Kaghan, and Owen-Smith all focus on technology transfer activities and offices, although at different levels of analysis. These offices are another institutional nexus for UIRR activities. Their existence is a fascinating case of what Morrill (2000) calls "interstitial emergence," that is, the emergence of organizational units (or firms) and professions to bridge the gap between institutional sectors. As "boundary organizations" (cf. Guston 1999) these units protect the boundaries between science and industry, even as they cross them.

It is the context of organizational theory that helps Hackett examine examines university institutional and value changes to help explain the consequences of changing characteristics of UIRRs in the chapter *Organizational Perspectives on University-Industry Research Relations.* He also discusses some of the problems from UIRR efforts, including confused lines of authority and responsibility, changes in evaluation procedures and priorities for faculty, complication of collegial relations based on allocation procedures for research support, and erosion of traditional academic values. His framework for studying UIRRs informs many of the remaining chapters of the book. He reviews resource dependency, institutionalist organizational analysis (Powell and DiMaggio 1991), and technocratic models of organizational activities and change. He suggests that UIRR development in a university is a signal of an emerging technocratic organizational form, although resource dependency and institutionalist theories help to explain the rise and legitimacy of their developments.

One particularly interesting organizational change, noted by Hackett, is the extremely rapid spread and ambivalent effectiveness of technology transfer offices (OTTs). The chapters by McCray, Kaghan, and, in part, Owen-Smith help to examine the functioning of OTTs. These are interesting units. Fligstein's (1987) account of the rise of finance personnel suggests that one mechanism for institutional isomorphism is the circulation of personnel. Powell and DiMaggio (1991) similarly suggest that professionalization helps to provide for the circulation of practices and organizational forms to address the question of the isomorphism of organizations within a given sector.

McCray examines OTT work ethnographically, illustrating the various perspectives that staff, faculty, and administrators have of this boundary organization. The work of the OTT is largely invisible. In acting as agents (Guston

1999), the OTT personnel on the one hand maintain the boundaries between academic work and industry, seeking to protect the interests of the different parties to intellectual property arrangements. On the other hand, they see themselves as educating the faculty and changing faculty behavior and beliefs about appropriate industry interactions. Kaghan examines OTT work ethnographically but with a broader unit of analysis. He looks at the larger culture of OTT professionals and the emerging practices, norms, and networks that provide a mechanism for the ideas, values, and practices of intellectual property to circulate among different institutions.

Owen-Smith looks at the role of intellectual property capture activities, namely patenting and licensing, and at OTT office expansion in relation to institutional stratification. Others (Gibbons et al. 1994; Bowie 1994) note that internal stratification is one of the problems of commercialization and suggest that interuniversity stratification may increase, yet little empirical data has been analysed to test the systemic effects of commercialization activities. Stratification may be linked to the systemic and internal differentiation that Clarke (1995) suggests is an inevitable outcome of the tensions between research, elite education, and mass education that contemporary universities embody. However, Owen-Smith's study suggests that there is an institutional version of the cumulative advantage that accrues to individuals (the Matthew Effect [Merton, 1973]). Thus the late adoption of OTT offices will likely do little for an institution that lags behind in reaping the rewards of university commercialization.

*Science as a Vocation in the 1990s: The Changing Organizational Culture of Academic Science*, also by Hackett, is an analysis of the effects of changes in academic science on the work and careers of young scientists. Based on Weber's (1946) classic treatment of "Science as a Vocation," Hackett asks many of the same questions about the quality of intellectual and personal lives of young scholars, given institutional change, changing resource "ecologies" and new pressures for efficiency and industry-relevant research. Hackett concludes that it is only with some trepidation that one might recommend a career in science to young scholars.

Several studies presented here use the research center as the primary unit of analysis and model of UIRR. Undergraduate educational experiences were also examined more broadly across the context of the university. *Building Labs and Building Lives* is a detailed ethnographic report of research laboratories and the ethics and values issues, that arise in the context of industrial and government sponsored or collaborative research projects. The focus is on students and education, in documenting structural changes in the university environment driving, and affected by, UIRR interactions. The first sections of this chapter look at values in undergraduate education, the later at graduate education and intellectual property concerns. Questions about the degradation of intellectual labor—the treatment of students as both machines for the production of knowledge, and as products of educational programs—are evidenced in the ethnographic data, information that is unavailable from other methods of considering values in university life.

In *Industry, Academe, and the Values of Undergraduate Engineers*, Hackett, Schneider, and Croissant argue that so-called cooperative education in engineering, where undergraduate students work in industrial jobs for periods of roughly six months, benefits both students and their employers. This is a largely unexamined form of univeristy-industry relations, and although it is focussed on education rather than research, the students may conduct research in the context of their internships. The attitudes of students involved in coop programs were compared with those of students undertaking more traditional academic research activities to see if the differing educational experiences had an outcome on value orientations. Mindful of the problems of student program selection based on pre-existing value orientations, the results suggest that students are indeed influenced by their educational experiences in industry and on-campus research activities. Students doing research valued the intrinsic rewards of education, expressed more interest in politics, and valued peer recognition more highly than students undertaking coop work. They were more likely than the industry-based students to state that one of their goals from an education was to become more well-rounded or simply better educated. In addition, cooperative education may have interesting effects on the reproduction of social class distinctions and achievement among engineering students. Similarly, cooperative education could be considered a variety of "school-to-work" curricula, which have generated some controversy. While promoting integration of curricula and expected needs for future workers, school-to-work programs also shift the curriculum to narrow interests and promote instrumental values in education. More generally, this research raises many questions, such as, do coops represent the privilege of those who can afford to leave campus, while internships reflect the resource bases of those who can afford to work for free? Or are internships and cooperative education programs opportunities for advancement for relatively disadvantaged students who can then obtain valuable experience and network contacts that are otherwise unavailable? Of course it is anticipated that the programs do both, but to whom do the relative advantages accrue most readily? This is one of the places where the call for additional research is most pressing.

Johnson presents a philosophical analysis of university faculty as professionals who balance a complex array of demands and normative considerations. Her chapter *Conflicts of Interest and Industry-Funded Research: Chasing Norms for Professional Practice in the Academy* considers faculty as professionals. She echoes Bowie's (1994) articulation of the problems of conflict of commitment and conflict of interest. Johnson specifically engages professional ethics, which allows her to argue that there are a number of principles and theories that can clarify the values conflicts—above and beyond the traditional normative descriptions of academic science—in faculty relations to students, sponsors, their departments and institutions, and the general public. Industrial funding of the professoriate exploits weaknesses in university faculty as a professional group, and the changing roles of professors requires more attention than they have thus far received.

Newcomer presents a detailed case study of the complex negotiations surrounding a research grant in *Your Space or Mine? Organizational Interactions and the Development of a Two-Arm Robotic Testbed*. He examines the intellectual, organizational, and educational changes that accompanied a grant and the reorganization of research groups around a specific device. The directed research agenda of the sponsor, NASA, focussed research efforts on a particular technological "instrumentality," to use Derek de Solla Price's term (1984), which became the center of conflict and a source of reorganizations of research and interactions among students and faculty. Differences in priorities, expected timing and output, and concrete research goals between the sponsors and research faculty contributed not only to an unsuccessful project but to the erosion of intellectual and resource autonomy for the faculty.

In a similar case study, *Pervasive Influence: Intellectual Property, Industrial History, and University Science*, Kleinman demonstrates in this particular case how in the agricultural sciences the role of industry provides important background assumptions that shape the conduct of science. Intellectual property concerns both constrain and enable laboratory activities and shape the direction and scope of research. Although this is emphasized elsewhere (Kleinman 1998), Kleinman's work also points out some of the mechanisms by which the world "external" to laboratories is always present and shaping scientific work, including such activities as problem formulation and prioritization, interpretation of data, and selection of research materials.

The "privileged position of business" in framing the assumptions of science is not new. For example, Clark (1997), writing about the field of occupational health sciences and regulation, notes that because universities receive grants from industry, they are "unlikely allies of labor" (178). Thus, medical science was slow to legitimate the occupational hazards of radium exposure in the 1920s. Kevles (1987), in his account of 1930s Marxist and humanist critiques of funding agendas for science, also reminds us that "if the choice of academic research subjects hinged at all upon the propensity of philanthropic foundations to fund noncontroversial studies, it also depended on criteria internal to professional disciplines" (246). If our cases illustrate that the internal criteria are increasingly convergent with business interests, however, where might we find the appropriate forms of disinterestedness and organized skepticism that would help to provide accountability for scientific and technological inquiry?

Since expert witnesses always testify on behalf of the side paying them (Clark 1997, 147), and researchers largely follow the resources available to them, even as they shape those resources, we should also consider some epistemic questions that emerge from these case studies. What happens to the objectivity of knowledge produced in institutions with low autonomy from the state and from industry? Or produced by researchers for whom curiosity and profit are no longer buffered by the norm of disinterestedness? As knowledge is always contested and contingent, our goal is not to retreat to a mythically pure notion of objectivity. As Daston (1992) notes, "our usage of the word *objectivity* is hopelessly but revealingly confused. It refers at once to metaphysics, to methods, and to morals" (597). At the least, we want to encourage critical discourse on the institutional bases for producing reliable, legitimate, and socially optimal knowledge (Bowie 1994; Fassin 1991). Likewise, our goal is not to retreat to the basic/applied science distinction or the science/technology split. Taking a page from Dasgupta and David (1987), the primary distinction between science and technology has to do with patterns of information disclosure and audience, not with the activities (laboratory work is laboratory work) or the presumed depth of theory. Packer and Webster (1996) argue that the shift toward patenting presents challenges to working scientists and technologists to negotiate distinct social worlds of credibility. What might changes of audience and decreased disclosure caused by commercialization and intellectual property protection activities mean for the perceived reliability and legitimacy of knowledge claims coming from universities?

The last chapter in our volume, *The Gloves Come Off: Shattered Alliances in Science and Technology Studies*, by Winner, is a statement about controversies over the role of critique in contemporary analyses of science and technology. The possibilities of critical inquiry into the social relations of science and technology are threatened by reactions from scientific communities, administrative apologists, business interests, and from both progressive and conservative commentators and scholars. In the immediate context, a conflict over the place of critique in academic institutions has been part of the evolution of this volume. At one point a center director protested our interpretations of center activities, especially when reported in draft form to our funding sponsor, the National Science Foundation. He wanted to evaluate our findings before we reported them. In his memorandum, he was concerned, along with matters of interpretation, with our publishing surcharge rates for research activities and our misidentifying employees of the Center as faculty of the Institute. We appreciated the opportunity for clarification, but those mistakes, in fact, illustrate our major points. What are the distinctions between private and public sectors in the context of business or state interactions with universities? Who makes these distinctions, are they legitimate, and why? Occupational categories at academic institutions are increasingly unstable. When are faculty not faculty? We have rapidly growing groups of quasi-faculty in the form of adjunct professors, non-Ph.D. instructors, and academic or classified professional staff who do the work of faculty in conducting research, teaching, and advising students.[2] How can, should, or will we hold adjunct faculty, or staff working as contracted employees, accountable as professionals and educators?

Since the center director has left his position at the Institute, we have not had to face further institutional conflicts over the reporting of this research. Nonetheless, the potential was there for a sizable controversy, perhaps impeding publication of a work critical of policies and practices at our own institutions. In a larger context, this work has since become more relevant as the salvos of "the science wars," larger attacks on the role of universities and intellectuals in public life, and emerging technological interventions in education (virtual and on-line education) have intensified.

In general, the decreasing resource autonomy of academic institutions and the rise of agendas, such as competitiveness or other state-initiated metrics of utility or productivity, bode poorly for the continuation of a tradition of universities as institutions sustaining a relatively autonomous intellectual sphere and thus acting as a source of social and cultural critique. However, we also know that universities are greedy institutions, and the costs of higher education rise far more quickly than the cost of living:

> The university shamelessly promised everything to everyone and charged so much that prospective students tended to believe the promises. While a state university, unlike an Ivy League institution, did not promise membership in the ruling class, the university, over the years, had made serious noises to all sorts of constituencies: Students would find good jobs, the state would see a return on its educational investment, businesses could harvest enthusiastic and well-trained workers by the hundreds, theory and technology would break through limits as old as the human race (and some lucky person would get to patent the breakthroughs). At the very least, students could expect to think true, beautiful, and profound thoughts, and thereafter live better lives. At the very very least, students could expect to slip the parental traces, get drunk, get high, have sex, seek passion, taste freedom and irresponsibility surrounded by the best facilities that money could buy. Its limits expanding at the speed of light, the university could teach a kid, male or female, to do anything from reading a poem to turning protein molecules into digital memory, from brewing beer to reinterpreting his or her entire past. . . . The university had become, more than anything, a vast network of interlocking wishes, some of them modest, some of them impossible, many of them conflicting, many of them complementary. (Smiley 1995, 386)

Thus all of the contributors to this volume are interested in values, such as accountability, and larger questions about the roles of science and technology in a democracy. We certainly do not aspire to rebuilding the walls of the ivory tower or to retreating to some mythic golden age of "pure science."[3] Nonetheless the questions of accountability have not been asked with sufficient breadth or depth. We would rather not wait several generations to ask the questions about the commercialization of technoscience that Leslie (1993) asks about its militarization:

> The full costs of mortaging the nation's high technology policy to the Pentagon can only be measured by the lost opportunities to have done things differently. No one now can go back to the beginning of the Cold War and follow those paths not taken. No one can assert with any confidence exactly where a science and engineering driven by other assumptions and priorities would have taken us. (256)

Rather than assume that economic competitiveness, privatization, and the transformations of universities through the emergence of new UIRRs will benefit science, technology, working scientists, various publics, or the students whom we aspire to educate, we thought we should interrogate those assumptions more directly. The following chapters are our attempts from various disciplines and methods to do so.

## NOTES

We would like to thank the authors for their enthusiasm, or at least forbearance, in the preparation of this volume. It has been a long time coming, but we all realized that too much had been invested and too much learned from each of the individual studies to simply abandon this collection. Special thanks are due to Dorothyanne Peltz, who copyedited and prepared the overall manuscript with great skill and even greater patience.

1. See the special issue of *Social Epistemology* 12(1), (January–March 1998) for essays on this topic.

2. Rhoades (1998) extensively studied these issues, particularly in the context of of faculty unionization.

3. As Gieryn (1999) notes: "The sociological question is not whether or not science is really pure or impure or both, but rather how its borders and territories are flexibly and discursively mapped in pursuit of some observed or inferred ambition, and with what consequences and for whom" (23).

# Abbreviations

| | |
|---|---|
| CBT | Center for Building Things (Pseudonym) |
| CCMS | Center for Composites, Materials, and Structures |
| CRS | Center for Research on Stuff (Pseudonym) |
| DARPA | Defense Advanced Projects Research Agency |
| DOD | Department of Defense |
| DOE | Department of Energy |
| IP | Intellectual Property |
| LA | Licensing Associate |
| NASA | National Aeronautics and Space Administration |
| NIH | National Institutes of Health |
| NRC | National Research Council |
| NSF | National Science Foundation |
| ONR | Office of Naval Research |
| ORU | Organized Research Units |
| OTT | Office of Technology Transfer |
| R&D | Research and Development |
| R1 | Research One Universities |
| RAL | Robotics and Automation Laboratory |
| S&E | Science and Engineering |
| SRC | Space Research Center |
| STS | Science and Technology Studies |
| TRC | Technical Review Committee |
| TT | Technology Transfer |
| UI(R)R | University-Industry (Research) Relations |
| USDA | U.S. Department of Agriculture |
| USPTO | U.S. Patent and Trademark Office |
| WARF | Wisconsin Alumni Research Foundation |
| WRF | Washington Research Foundation |

# 1

# Organizational Perspectives on University-Industry Research Relations

## EDWARD J. HACKETT

Academic science is undergoing a profound transformation. Faculty roles and working conditions are changing (Hackett 1990), entrepreneurial behavior is encouraged (Blumenthal, Gluck, and Louis 1986; Blumenthal 1996; Etzkowitz 1983; Louis et al. 1989), the peer review system has undergone continual scrutiny (Chubin and Hackett 1990; U.S. General Accounting Office 1986; Committee on Criteria for Federal Support for Research and Development 1995), universities eagerly partake of the pork barrel, sometimes assisted by professional lobbyists (Chubin and Hackett 1990; Cordes 1991, 1998), and for-profit institutions of higher education are rising in prominence. In all, the social contract between universities and other parts of society has been rewritten (Gibbons et al. 1994; Remington 1988; Slaughter 1988; Slaughter and Leslie 1997; Smith 1990). Of all such changes, the growth and strengthening of university-industry research relations (UIRRs) is one of the most profound and revealing organizational changes to influence United States universities. There are several reasons for this judgment.

First, UIRRs embody universities' increased reliance on private sector resources, a reliance accompanied by expectations that a tangible product or service will be produced in return. More than a simple contractual arrangement, such relationships often entail significant changes in the organization and functioning of parts of the university. Technology transfer offices, patent attorneys, institutional review boards, and the like are structural accommodations to the demands of outside organizations for greater accountability and more "businesslike" practices. Educational programs and students' interests are also shaped by the increasingly tight connection between academe and industry.

Second, since 1980 there has been a rise in the number and power of non-academic professionals on university payrolls, including managers, planners, lawyers, and accountants. Such professionals often broker relationships between universities and businesses, perhaps by marketing a university's services, negotiating contracts, or accounting for expenditures. They also represent the university's response to calls for them to adopt more "businesslike" ways. There has been a long-term correlative rise in the professionalization of college administrators, complete with specialized academic credentials and memberships in professional societies, supplanting "amateur" administrators who had moved into administrative posts from faculty ranks.

Third, faculty have been increasingly driven to seek external support for their teaching and research, but federal support for higher education has not kept pace with such demands, so industrial sources have become an attractive alternative. The imposition of such "market discipline" on the expenditure of faculty time may have placed exchanges among faculty, staff, students, and administrators on a more instrumental basis. Internal accounting systems, such as treating academic departments as cost centers, make visible and measurable exchanges that had been intangible and implicit.

University-industry research relations have been described (Peters and Fusfeld 1982), evaluated, encouraged and lamented (in countless popular articles and speeches by politicians and university presidents), and folded into clever theoretical frameworks (Etzkowitz and Leydesdorff 1997). Historical studies have placed UIRRs in the context of universities' strategies for visibility and self-aggrandizement (Geiger 1986; Leslie 1987; Weiner 1986). Some helpful souls have even provided how-to lessons for those attempting to build such connections (e.g., Powers 1988).

What has not yet been done is to examine UIRRs from an organizational perspective, looking for systematic insights into their origins, inner workings, and likely trajectories. Complementarily, UIRRs offer a strategic but somewhat neglected site for studying new organizational forms and processes. This paper is a modest effort to address both needs. Empirical illustrations will be drawn from a qualitative study of ethical and value issues posed by UIRRs. The study included an examination of center documents, observation of center meetings and other work, and interviews with center administrators, sponsors, members, affiliated faculty, and university officials.

## TWO CENTERS

Our research focused on two centers: the Center for Building Things (CBT) and the Center for Research on Stuff (CRS), located at Hilltop Tech, a northeastern engineering school. The CBT and CRS are different in many significant respects.

CBT was initiated by university administrators, whereas the CRS was created by faculty. The top-down versus bottom-up origins of the centers influenced their receptions on campus and the meaning faculty ascribed to participation. Some faculty thought the CBT did not belong in a university and that participation in it was a form of selling out to administrative or industrial interests, while the CRS was accepted as a larger version of appropriate faculty research activity.

In its infancy the CBT took on a wide assortment of small, focused jobs to build its research portfolio and credibility (becoming, as some said pejoratively, a "job-shop"). In contrast, the CRS initially was funded by a large grant from the U.S. Department of Defense, later supplemented by industrial partners and research sponsors. In effect, the CBT was industrial in origin, migrating to governmental support, while the CRS originated with government support, parlaying this investment into industrial participation. This foundational difference shaped the level and character of research activity in the two centers, influencing their time horizons, stability, and legitimacy.

At its inception the CBT committed in principle to limiting the amount of government research support it accepted to remain responsive to industrial needs. Indeed, in the minutes of a meeting of CBT's advisory board of industrial sponsors during the early 1980s, participating companies agreed that "government agencies would be allowed to support only projects of manufacturing application efforts and not projects for generic research" (official minutes of the meeting, 1981, 81). Significantly, near the end of the study period, when industrial research support became scarce, the CBT loosened this guideline and pursued contracts with military facilities and other government agencies. In contrast, the CRS made no such commitment and has been overwhelmingly supported by federal money. Private companies were invited to become members of the center, for $50,000 per annum, to get an early and close look at the fruits of this major federal research investment.

In structure the CBT resembled a business, with a director, associate directors, and clearly stated principles of organization, hierarchy, performance and accountability, whereas the CRS is putatively more collegial, informal, and amorphous, that is, more traditionally academic, both to the outside observer and in its self-perception.

The CBT has devised innovative project teams consisting of non-tenure-track staff (some with Ph.D.s, others graduate students, all with industrial experience) who are responsible for budgetary, contractual, and administrative matters, working in tandem with regular faculty who provide technical expertise, research credibility, and academic responsibility and oversight. To oversimplify somewhat for clarity: staff provide money, faculty provide students. In contrast, the CRS has a more traditional faculty-researcher pattern, similar to what one might find in an NIH-funded program project, consisting of multiple, semi-independent investigators sharing a pool of resources, laboratory space, equipment, and assistance.

## THREE ORGANIZATIONAL PERSPECTIVES

Three theories of organizational behavior will be outlined, illustrated with case material, and used to explain the origins and operations of these research centers. I will briefly sketch the main principles of each perspective, assess its fit with the case material, and outline its research prospects and theoretical promise.

### Resource Dependence

The resource dependence perspective calls attention to the flows of resources (or direct exchanges) between a focal organization and the other organizations in its environment, emphasizing the instrumental exchanges that allow an organization to accomplish its objectives (Thompson 1967; Aldrich and Pfeffer 1976; Pfeffer and Salancik 1978; Tolbert 1985). In this view, organizations are open to their environments and in continual exchange with others to obtain needed resources. Organizational structures might change through the addition of offices, units, and procedures to insure a stable flow of resources and to cope with the problems and uncertainties that arise from changes among its trading partners (Tolbert 1985, 1). Environmental circumstances pose a variety of challenges, and organizations respond by actively developing strategies for managing their environments.[1]

In the case of university-industry research relations, declining enrollments and reduced federal support for research and infrastructure, in combination with U.S. economic problems in the mid-1980s, the enduring emphasis on high technology, and persistent (but controversial) claims of shortages of scientific and technical employees, changed the university's resource position and altered the prospects of various organizations in its environment (Slaughter and Leslie 1997). In response, universities sought and received greater amounts of industrial funding.

National data for the period give some indication of the level of dependence. From 1980 to 1987, industrial support for academic research rose by 133 percent (in constant dollars), accounting for 6.4 percent of all academic research expenditures, up from 3.9 percent (National Science Board 1989). By 1995 industry provided 6.9 percent of academic research support (Appendix, National Science Board 1996, 104–106).

A closer look at the data shows a stronger pattern of potential dependence. Among the top 200 research universities, those with lesser amounts of total research support tend to receive higher fractions of support from industry (National Science Board 1989). While universities with substantial federal research support often have substantial industrial support as well, those that have lower levels of federal support receive a higher fraction of their support from industry. In general, industrial funds amounted to about 5.5 percent of the research support for the seventy-five schools with the most total research funding, about 7.5 percent for the next seventy-five schools, and about 10.5 percent for the final fifty. In fiscal year 1987, Hilltop Tech received about a third of its research support from industrial sources (National Science Board 1989, 305).

The resource dependence perspective, applied to the circumstances of UIRRs, highlights the structural consequences of environmental change and the conscious strategies and tactics leaders employ to guide organizations through turbulent times. For example, to obtain government grants and contracts, universities might add offices that provide appropriate guarantees of due process and accountability, and perhaps create liaison (boundary-spanner) positions as well. Universities added vice presidents for government relations (or other positions with similar titles, often hired directly from the upper levels of federal agencies) to strengthen their connection with local, state, and federal governments. In response to the growing importance of industrial support, universities have added industrial liaison organizations (such as centers, institutes, industrial affiliates programs, research parks, and the like). Some universities rapidly enlarged their technology transfer offices, not only to encourage and to profit from faculty inventions, but also to build a bridge to the private sector. Candidates with high-level industrial experience were considered presidential timber. All these would be predictable organizational consequences of resource dependence.

This perspective also highlights the cooptive character of organization-environment relations: organizational leaders do not simply accept their environments as they find them but attempt to gain control in various ways. University presidents attempt this through speeches and press releases, through service on national panels and commissions, and through appearances at professional meetings. Universities also endeavor to change national policy through their states' elected representatives and through the services of professional lobbyists. Reciprocally, the companies that exchange with universities also try to shape research agendas, graduate and undergraduate education, student career choices, and general university policies and practices. So each partner is attempting to co-opt the other in some respect.

The resource dependence perspective has much appeal, because it focuses squarely on what the university has at stake in UIRRs: a sorely-needed source of additional support. While academic reliance on industrial support has remained fairly high, and Hilltop Tech has sustained its commitment to industrial involvement, opinions are divided about the lasting effects of such changes on the structure and operations of the university. Some contend that universities are involved with industry for purely economic reasons, not from any deeper change of mission or new-found commitment to national economic competitiveness, and that this involvement is being handled at arm's length. In the words of one official at Hilltop Tech, "a center exists to be an interface with industry. . . . It's a marketing tool of an academic institution that provides an input terminal for all kinds of resources from industry to flow into the institution."

But others contend that UIRRs represent a new commitment, suggesting that the influx of industrial support will bring about lasting cultural change. In the words of George M. Low (1980), a leading spokesman for UIRRs: "If we accept the premise that some of the problems that have led to our loss of competitiveness can be solved by enticing larger numbers of our best engineering graduates into design and manufacturing, then we can begin to provide solutions by strengthening the linkages between industry and universities" (3). In this view UIRRs are more than "input terminals" for resources, representing instead the vanguard of thoroughgoing and lasting change.

However consequential the change may be, some inconvenient facts weaken the resource dependency argument. Over the period under consideration, industrial spending for research and development on university campuses rose from $334 million in 1980, through $920 million in 1989, to $1.165 billion in 1995 (all in constant 1987 dollars), accounting for only a small fraction of the university research budget and hardly a level of resource dependence likely to drive lasting structural change. Indeed, industry provided 6.2 percent of academic research expenditures in 1960, raising the historical question of why and how the organizational response occurred in the mid- to late-1980s rather than in the early 1960s (Appendix, National Science Board 1996, 104–106). Finally, the resource dependence perspective focuses substantially on instrumental activities, direct exchanges, and organizational structure to the neglect of symbolic activities, indirect relationships, and cultural changes. Yet much social activity is not instrumental, and some organizations appear to be influenced by others, even in the absence of direct exchange. Perhaps most importantly, much of the rhetoric associated with UIRRs tries to reconcile these new activities with the traditional academic mission and values. Such efforts at cultural housekeeping call attention to the possibility that a significant disturbance of mission and values is taking place. Perhaps the most trenchant questions about UIRRs have to do with the changing norms and values of academe—inherently cultural matters—and resource dependence is generally silent on such matters.

Many of the shortcomings of the resource dependence perspective are addressed by the "new institutional theory" of organizations, which developed during the late 1980s and early 1990s.

## Perspective II: Institutional Theory

Institutional theory begins with the observation that there are striking similarities in structure and culture among organizations within an organizational field (which includes a focal organization, its customers, resource providers, and competitors) then explains such similarities by examining how organizations acquire and maintain legitimacy. Importantly, organizations in a field need not engage in direct exchange relationships to influence one another; they need only take account of one another, even at a distance. Unlike resource dependence, which focuses on patterns of exchange and conscious decisions, institutional theory recognizes the influences of symbolic relationships that may involve no exchange and that may act despite actors' intentions or awareness.

Institutional theory is especially concerned with the ways societal understandings and expectations about appropriate structure and behavior bring about conformity in a population of organizations (Tolbert 1985, 1). By conforming to these normative expectations, the theory holds, organizations gain legitimacy (and, therefore, access to essential resources) and thus increase their chances of survival. Perhaps stated too simply, resource dependence theory is mainly concerned with structural relationships, particularly exchange, dependence, and purposeful leadership, whereas institutional theories concentrate on symbolic interactions, particularly the attainment and maintenance of legitimacy. These explanations are potentially complementary, as Tolbert (1985) has shown, but there is much negotiation ahead before this can be achieved (Carroll and Hannan 1989a,b; Zucker 1989; Scott 1995).

The institutional perspective has two main advantages: (1) it focuses attention on legitimation and on the cultural beliefs and values that shape the organization and are served by its work; (2) it raises the possibility that things may not be as they seem, that the cultural beliefs and values espoused by an organization and its members may be observed only through ritual and symbolism, while the main business of the organization goes on undisturbed (Meyer and Rowan 1977). We are led to ask whether universities are seeking new grounds of legitimacy and, if so, why and how? How do universities reconcile their new activities with their traditional mission to educate and to produce new knowledge? And we are cautioned to view apparent consequences with some skepticism. But institutional theory also has several significant weaknesses.

In the first instance, institutional theory is short on mechanism. Exactly how do institutional forces operate on an organization? Coercive, mimetic, and normative forces are the answer proposed in DiMaggio and Powell's (1983) classic paper, but that still leaves much unexplained. How is coercion exercised? Who effects the copycat behavior of mimesis, deciding which organization to emulate, which features to copy, and how to bring about the transformation? Why and how do professionals enter organizations, bringing with them the norms that will change the place? And how do they carry out the changes? Facing such questions, one again wants to integrate the almost mystical workings of institutional forces with the undeniable reality that the need for resources shapes behavior and that at times powerful actors will take direct action to achieve their aims.

In the specific case of Hilltop Tech, the university president stated publicly that Hilltop's strategic decision to seek industrial connections—the strategy that gave birth to the CBT and other centers—was in large part motivated by a conscious desire to differentiate Hilltop Tech from other universities so it would not be seen as "just another Boston Tech." Thus its receptivity to industrial support was based partly on need for resources and partly on the symbolic or reputational advantage of a distinct identity.

Second, empirical studies informed by institutional theory have mainly focused on structural features of organizations, although its central idea, legitimacy, has more to do with such cultural considerations as the values, norms, rituals, and customary practices that orient and guide behavior. Further, such cultural phenomena are difficult to observe and interpret. For example, institutional forces may alter the workings of existing structures or may change the meanings assigned to such workings, rather than giving rise to new or revised structures. Both repurposing and reinterpretation would be difficult to observe through usual research techniques. A case in point: as universities try to conform to the expectations of private industry, the character of mentorships and research collaborations may change, although their structure—their constitution and outward appearance—may not be visibly altered. When we consider UIRRs, our eye is drawn first to the obvious structural changes: the centers and institutes that have arisen in response to the prospects of industrial support. But we must not ignore the consequences of such changes for academic culture and for the values espoused and embodied by students, faculty, and administrators.

Institutional theory also stops short of discussing the consequences of the quest for legitimacy on the substance of an organization's work, that is, on the content and character of its products. After all, when we think about academic work, we quickly wonder how such arrangements affect scientists' and engineers' teaching and research, and students' learning and values. Although students and faculty may continue to write and publish papers, to teach and take courses, the content, audience, purposes, and effects of those activities may change.

Institutional theory invites us to ask a larger question that bears on the content and character of academic work. How are fields of science created to legitimate activities undertaken on university campuses? Among such fields are agricultural science, biotechnology, computer science, manufacturing, decision sciences, and materials science. These are not traditional academic fields, but as they become legitimate academic pursuits, complete with journals, conferences, professional societies, departments, and degree programs, the activity becomes properly academic. The bootstrapping is mutual. The nascent discipline gains legitimacy because its activities are undertaken in a university, while the university gains legitimacy because the work in which it is engaged is part of a recognized discipline. And, of course, industry gains by impressing its interests on both the university and the discipline.

This example from the study illustrates how interaction with industry, including an infusion of money, can shape the content and legitimacy of core academic activities. In the words of a faculty member at Hilltop:

> A few years ago IBM gave $2 million here for the master's curriculum development, which I thought was one of the silliest things. . . . I couldn't believe it. There were five schools in the United States that got $2 million each. . . . But, ask yourself, what did IBM get for $10 million? My answer is that, for $10 million, IBM legitimized manufacturing as an academic discipline. There were something like fifty or sixty proposals submitted [for the five awards of $2 million each]. . . . Now, of the fifty or sixty originally submitted, something like 80 or 90 percent of them have manufacturing programs in place anyway. So, for $10 million they legitimized manufacturing as an academic discipline. It is now a viable career path for a student. . . . And the same thing with the faculty. You know, faculty never do research in manufacturing but, by putting this heavy chunk of money in there. . . .

Institutional theory connects changes within an organization to an overarching process of bureaucratization, a Weberian rationalization of the world. Yet many UIRRs (and other organizations) are not especially bureaucratic in form or operation. Technical and administrative functions are blended; lines of authority are blurred; tenured faculty are essential to the endeavor but notoriously resistant to hierarchical control; boundaries are permeable (with resident engineers studying on campus while paid by their companies and "co-op" students leaving campus for paid work in companies); team memberships intersect with staff working across project lines; the distinction between the public and private person is blurred through intellectual property and nondisclosure agreements.

Universities may be more complicated organizations in a more complicated environment than many of the organizations considered by institutional theorists. Universities have diverse goals; strong and inconsistent values; varied stakeholders; products that are symbolic and tangible, human and nonhuman; and occupy a shifting place in society. They are influenced as well by the distinctive cultures and purposes of the various disciplines represented within—in all, a form of organization that is neither very rational nor very bureaucratic.

## Perspective III: The Technocratic Organization

UIRRs may represent a new organizational form, such as the "technocratic organization" proposed by Heydebrand (1989). He argues that new forms of organization are emerging as a consequence of the transition from industrial capitalism to postindustrial capitalism. These technocratic organizations are

> Small or are small subunits in larger organizations. . . . They are staffed by spe-
> cialists, profes sionals, and experts who work in an organic, decentralized
> structure of project teams, task forces, and relatively autonomous groups.
> There is little emphasis on a formal division of labor and managerial hierar-
> chy, with managerial and technical functions overlapping to some extent. The
> loosely coupled organizational structure is frequently reorganized and cen-
> trifugal. Such conditions require new methods of fostering social cohesion,
> such as informal interpersonal relations, clanlike norms and practices, and the
> creation of a corporate culture.
>
> In sum, the new organizational forms are postbureaucratic in that they
> move away from formal rationality, a fixed hierarchy and division of labor, for-
> mal procedural specifications of work relations apart from computer software
> and rigid norms of formal interaction and deference (337).

Heydebrand's claim, then, is that large organizations may respond to changes in their political and economic environment by creating subunits with these distinctive properties. Unlike the more thoroughgoing organizational changes implied by resource dependence and institutional theories, this is a partial or segmented adaptation. That is, the organization responds to changed circum-stances by creating a part of itself with characteristics very different from its core, a part that can both adapt to the new circumstances while simultaneously serving as a buffer for the organization's core activities and values.[2] Of greatest applicability to the case of UIRRs is the idea of a postbureaucratic organiza-tion grounded in "technocratic rationality," which

> Can be seen as a fusion of formal and substantive rationality and as a transi-
> tional stage that opens the possibility of development of other forms, for
> example, democratic ones, since it tends to undermine or remove two of the
> most formidable obstacles to organizational democracy: bureaucracy and pro-
> fessional dominance. (Heydebrand 1989, 344)

Heydebrand goes on to characterize technocratic organizations along sev-eral dimensions, which are combined and simplified for presentation here.

*Informalism.* Technocratic organizations, Heydebrand (1989) argues, have a dis-tinct, informal rationality

> That differs from both the formal rationality of bureaucracy and the substan-
> tive rationality of professional control and political power. In work and author-

ity relations, an orientation toward problem solving, results, the weighing of interests supersedes bureaucratic rule orientation and behavioral correctness in following procedural norms . . . increasing the flexibility of social structures and making them amenable to new forms of indirect and internalized control, including cultural and ideological control. . . . [S]pecial interests and prerogatives such as are typically defended by occupational groups and professional elites are undermined, weakened, or integrated into systemic arrangements. (344–345)

The UIRRs we observed, particularly the CBT, both embody these objectives and are themselves compromises designed to solve industry's practical problems and need for specially trained workers while allowing the university to accept industrial support without explicitly changing its departmental structure and mission. In a sense, the process is similar to that predicted by the resource dependency framework, except that the dependence-induced change is limited to only a part of the whole organization. UIRRs also effectively compromise both the authority of the university administration (because the technical aspects of their research are beyond the grasp of most administrators) and the professional authority of the faculty (because projects and centers are organized in ways that weaken the autonomy of faculty participants).

Unlike classic, investigator-initiated research awards, a substantial amount of UIRR funding is not designated for specific researchers. Rather, it is up to the center administration to allocate the support to investigators within the center. As one respondent told us,

> [W]hat the government basically has done is put some of the management that they would have had before at the university level. So where they might have let twenty contracts before, they now give one big contract and let you do the management on it at the university level and handle those problems. But you aren't really set up to do that. . . . and the amount of work goes up astronomically. [CRS materials scientist]

Such management and allocation issues were a source of sharp dispute in both centers, with some investigators feeling ill-used by those who made the decisions. Allegations of unfairness, even deception, were made.

Of course, scientists whose research proposals have been declined by any potential sponsor always feel somewhat ill-used; after all, they typically worked quite hard on the proposal and counted on the funding for equipment, supplies, and support. But in this case the decision makers are housed within the complainant's university, perhaps within the same center or department, and are encountered in everyday life on campus. The allocation process may not be specified in much detail; in fact, it may be unstated or arbitrary, and there may be no appeal or other recourse. Finally, the absence of an established peer review system, with its attendant safeguards and legitimacy, makes unfavorable decisions harder to accept. In all, internal management of omnibus grants and contracts may complicate collegial relations and encourage the formation of cliques and other alliances.

One untenured assistant professor affiliated with the CBT complained bitterly and at length about how industrial resources that came into the center were distributed among faculty. The CBT, in his view, "has been playing a bait-and-switch game with our industrial sponsors and with [my school]. They have been using our [that is, his group's] . . . talent and expertise and knowledge to help sell their programs and . . . they keep the money for themselves." And "the money" in this case amounted to more than $100,000 per year.

When asked how allocation decisions were made, whether by executive decision, in a committee, or by some other process, the professor replied, "It's not clear how they're made. I would bet that it's senior . . . faculty and administrators that'll make the decision. . . . It's probably best that it be somewhat informal or flexible. . . . I mean, if a rigid set of policy guidelines was to say how we are to allocate these funds, probably it wouldn't work, either."

But the informal mechanism didn't provide the level of support for his group that he felt it needed and that the industrial sponsor intended it to have. Worse, "[w]e didn't have the money to [fund the group at the intended level]. The only way we could do that was to decrease someone else's funding, and for political reasons it was decided that. . . . I mean, I didn't feel comfortable as an untenured faculty member saying 'You got to cut these guys' funding in order to support my work'." Despite this reticence matters soon came to a head.

> Ultimately, in October, I ended up giving [CBT] leadership an ultimatum, which is "You either increase the funding or I'm gone, and our whole group is gone. We have waited for six months and you have not given us a firm commitment as to when the funding will be increased or to what level, and I'm not going to wait any longer." And the decision was made to increase it, starting in January. Frankly, that nine months was pretty important for me because it comes at kind of the middle part of my tenure process. . . . January's almost a little too late.

Informal methods of resource allocation can generate a lot of heat and some undesirable side effects. But solidarity is strong enough that the professor, despite feeling wronged, did not seek redress outside the center, neither from university administrators nor from industrial sponsors. And that solidarity is not structural, based on a clear hierarchy and clear lines of authority and responsibility, but appears to be normative, a part of the organizational culture.

UIRRs have also created new and confused lines of authority and responsibility. For example, the director of the CRS is a member of an academic department chaired by a member of his center. Also in that department and in that center is the Principal Investigator of a very large federal project—the largest by far within the center or department—which includes as co-PIs the center director and the department chair. Thus each reports to the others in some capacity, and each relies on the others for some form of support. While this sort of cross-cutting authority can be the source of interdependence and solidarity, it can also generate a lot of friction and heat. And it has.

*Permeability.* Technocratic organizations are permeable in that boundaries between categories, concepts, social roles, and organizations are dissolved or circumvented. Abstract principles or concepts, particularly those that pose "either/or" choices, are recast as "both/and" possibilities. In this way technocratic organizations may be subversive agents of change, because they use existing ideas and categories, even ideas and categories that may seem inconsistent or contradictory, in new configurations. In doing so they create

> Weak, permeable boundaries between cognitive categories, areas of jurisdiction, and the framing of options, spheres of competence, hierarchical levels, and social roles. There is a distinct tendency toward the fusion of authority and knowledge, of managerial and expert functions. Formal distinctions among officials, and between officials and clients, tend to disappear, and client participation in decision making is encouraged up to a point. (Heydebrand 1989, 345–346)

Here is an example of how boundaries between conceptual categories and jurisdictions may be blurred. It is drawn from an early report of the CBT in which the center declared that long-accepted boundaries and distinctions could be set aside, "conventional wisdom" notwithstanding:

> *Conventional wisdom* says that research universities do not do *applied* research because it does not include doctoral-level research. The *reality* is that many applied research objectives do embody doctoral-level challenges. . . . [the CBT] has developed . . . the ability to concentrate on research that is both scholarly and meaningful to industry.
> *Conventional wisdom* says that universities promise a "best effort" rather than results. In *reality*, there is no reason that a university research center cannot provide a structured environment of creativity which provides management based on time, cost, and performance. In fact, there is a strong indication that time, cost, and performance management can enhance the likelihood of creative research, significant breakthroughs, and scholarly results. (CBT report, 1981 [emphases in original]).

These are profound statements, more so because they are made without evidence or authority. In effect, they assert that old distinctions between basic and applied research, and old expectations about the tension between freedom and accountability in research, simply are not appropriate in this new research organization. Moreover, the assertion is made in a tone that is almost charismatic: "it is written . . . , but I say to you. . . ." It is also noteworthy that the CBT is challenging the longstanding contrast between "best efforts" and "results" in academic contracting. When used in a contract between the university and a research sponsor, "best efforts" commits the university to a certain process of research without promising a particular outcome or product. In its document the CBT goes boldly—at least in principle—where most universities prefer not to tread: asserting its willingness to promise results and not merely the research equivalent of "the old college try." Further exemplifying the rationality of technocratic organizations, these assertions are made in pragmatic pursuit of resources and legitimacy.

Similarly, the research team arrangement (that combines technical and administrative responsibility) and the presence of project engineers on campus (and students taking cooperative employment for a time in industry) also attest to this permeability, this fusion of purpose and personnel across traditional boundaries.

The limits to permeability and the boundary work undertaken to maintain those limits offer a point of contrast that emphasizes this claim. One professor associated with CBT said this about his industrial sponsors:

> There's a little bit of a bait-and-switch that we can pull with these industrial sponsors. We tell them, you know, "You're going to have a chance to participate [in the research] and this is what we're going to do with it." [But] "Your role from my point of view is quasi-steering, quasi-advising, and probably you're advising [not steering] for the following reasons. . . . You're practitioners and you do it, but you've never been trained to do research on this topic, and it's hard for you to set priorities as to which segments of the research have to be done first, how the various segments of the research interrelate."

In this way sponsors are involved in research, but only to a point. And they are met halfway at that point by the ways research problems are chosen and configured. For this professor, the compromise is accomplished by seeking problems that offer both real-world relevance and academic rigor. Research problems are sought or constructed in ways that matter to sponsors and use variables that can be manipulated in the real world but that also are simple enough to be solved in an academically rigorous fashion. To increase the industrial relevance of their work, graduate students write their theses in two forms, one that incorporates real but confidential data obtained from the company and another that uses hypothetical data that are suitable for shelving in the college library.

*Flexibility.* Technocratic organizations have a flexible, open, malleable structure. Within organizations as tradition-bound as universities, they are islands of responsiveness and innovation. Indeed, technocratic organizations "place a premium on the exchangeability and expendability of subunits and on policies favoring a flexible, modular structure. This opens the possibility of experimentation, innovation, and the injection of new programs as well as the possibility of dropping obsolete or unprofitable ones" (Heydebrand 1989, 346). They also "Encourage the design and formation of external linkages with other organizations in their environment, including agencies of the state at local, state, and federal levels. The flexibility of networking strategies enhances the "organic," "responsive," and reflexive nature and openness of the organization vis a vis its external environment" (Heydebrand 1989, 346–347).

Not only do research centers and institutes provide university administrators with flexibility—it is far easier to close a center than it is to close a department—but they also serve as a buffer between the university's core activities and the turbulent environment. At a lower level of organization, programs and projects, and the staff who operate them, also provide center directors with flexibility.

UIRRS as technocratic organizations create new academic roles, especially roles designed to manage inconsistent demands or to span organizational boundaries. Such roles have been termed "academic marginals," although others have called them the "unfaculty" or "unequal peers" (Kruytbosch and Messinger 1968; Teich 1982). These people are fully-trained scientists, often with doctorates, who occupy a never-never land between postdoc and faculty member. At the CBT, marginals have been structured into a unique project management scheme. Each project is the responsibility of at least two people: a regular faculty member who provides intellectual leadership and academic legitimacy to the endeavor, and a project manager who is chiefly responsible for budget, reporting, technical aspects of the work, and deliverables. These project managers are a form of academic marginal. While they generally have masters degrees, an appropriate terminal degree for engineers doing such work, they occupy that vague territory between faculty and students. They hire students and can remove them from projects, even over the protests of faculty. Further, our fieldwork documented that they provided much of the day-to-day supervision of students, both for project and for academic work. But they cannot take formal responsibility or credit for students' work, and even find themselves donating time to teach or otherwise participate in academic affairs.

Despite such limitations, the liminal status of project managers makes them ideal boundary-spanners, and in such roles they provide an essential service to Hilltop Tech and to companies. As one CBT project manager explained: "Project managers spend 20 percent to 40 percent of their time talking to high-level people in industry. We serve a very large sales/public relations function that faculty don't have the resources or time to do."

Another facet of flexibility is exhibited in the complicated internal structure of UIRRs. For example, the CBT has given rise to three substantial centers that reside wholly or partly within it: one funded by a consortium of industries, a second supported by federal funds, a third paid for by the state. The state and federal centers are substantially devoted to technology transfer, whereas the industry center conducts research. Each part is a significant fraction of the whole.

At the aggregate level of the entire university, the creation and growth of centers, institutes, and technology parks on or near campuses is itself an indicator of flexibility. Similarly, offices within universities that promote government relations, technology transfer, economic development, and the like are structural embodiments of organizational priorities and commitments. Hilltop Tech added two vice presidents, one as a result of splitting the joint vice presidency for administration and finance because there got to be too much of both for one person to handle and the other a new vice-president for government relations. Creating such a position signals the propriety and importance of "government relations" to state and federal officials and to other universities.

UIRRs are richly joined to their environments through Industrial Steering Committees, resident engineers, industrial projects, students venturing off into cooperative employment, and the like. In one striking case, a company established at Hilltop Tech a project so sensitive that it occupied a locked laboratory with blacked-out windows, limited access, and secrecy rules restricting communication—embassy for research, industrial soil on a university campus. On the other side of the coin, within the CBT a federally funded center was established to assist in the transfer of technology from government laboratories to the private sector. To do this the center created a demonstration site (where company representatives can try out the technologies) and provided other forms of technical assistance. Eventually the demonstration center was spun off and became independent. A state-funded center, also focusing on technology transfer, grew out of the CBT and then provided seed money for some of its research efforts.

The most striking evidence of such interpenetrations comes from an early meeting of the founders and administrators of the CBT, which was held in the early 1980s. Staffing and recruitment were prominent topics, with the CBT director noting that:

> If Project Engineers can be transferred from sponsors [companies] to Hilltop Tech, as suggested, the sponsor could help with moving expenses and also in making up the salary differential that might occur. In return this Project Engineer could work on a project submitted by the sponsor and at the same time take courses at Hilltop Tech, which would increase his skills and make him technically more proficient on his return to the sponsor's facility. [official meeting minutes, 1981, 10]

The director further suggested that "During this recessionary period, in particular, founding companies could transfer some of their bright young engineers who are being laid off to Hilltop Tech. Suitable agreement could be made whereby these engineers did not lose continuity of service and could . . . return when the economic situation improved" [official meeting minutes, 1981, 10]. In the ensuing discussion companies suggested development of an exchange program, with faculty trading places for a time with company employees.

Another example of the intensity of organizational interpenetration, which arose at the same meeting, was a discussion of the "antitrust implications of a vendor and a purchaser both sitting on the advisory board of the CBT. Would the supplier on the advisory board have an advantage over other vendors of that association?" Members were worried that the competitive advantage they might gain by sitting together on the board would be illegal, preventing their doing business together.

Antitrust violations arising from connections among companies are certainly not novel, and employees changing employers during an economic recession is also not new. But UIRRs are a new site for such events. That possibility reflects the peculiar character of such organizations, and the fact that they do take place on campus reflects a new role for universities.

The CBT Engine offers a programmatic statement of the desired blending of the interests and purposes of various stakeholders in the center. Quoting CBT's 1981 annual report:

> The last year has epitomized the short-term development project, long-term basic, and applied research balance and cycle that the Center has identified as optimal for serving all our constituencies: industry; faculty; staff and students. Growing technical expertise and laboratory facility through a series of related projects occasionally suggests a longer-term, broader-scope, basic research program. Conversely, once basic research results start to emerge from programs, application-specific projects are frequently spawned. The result is to balance the Center, which will enable our continued march toward worldwide recognition for excellence.

Notice that everything is here, swirling along in fruitful synergy. Basic research, applied research, and development reside together in harmony, each begetting the other and all benefiting all concerned with no tradeoffs, no distinctions, and no differences in interests, values, or perspectives to complicate matters.

*Cultural solidarity.* Unlike bureaucracies, which are built upon a foundation of rational-legal authority, technocratic organizations "must seek to build a corporate culture. . . . which acts more like an umbrella for a variety of strategies and scenarios . . . and serves to counteract the centrifugal tendencies of decentralization, loose coupling, and weak, permeable boundaries" (Heydebrand 1989, 347).

Within UIRRs control is exercised by asserting the greater importance of the center *qua* center over the cross-cutting claims of individual disciplines, faculty members, and even the university itself. It is also sometimes accomplished by appealing to the center's constitutive mission and unique contribution to society. For example, the CBT's mission is "to serve our sponsors and the nation with innovative manufacturing education, research, and technology transfer necessary for their leadership in the international marketplace." By asserting this overarching goal, CBT leadership diminished the special claims and independence of faculty and their academic disciplines, subordinating them to the center's needs.

A distinctive organizational culture is also promoted within UIRRs. For example, stationery and report covers used by the CBT display the center's name and logo more prominently than the institution's. Tee shirts with a catchy slogan ("Manufacturing the future.") reinforce the message. The CRS achieved solidarity in a different manner, engaging in a protracted skirmish with the administration over space, resources, and renovations in such a fashion that created an us-versus-them mentality. In different ways, both centers promoted solidarity by setting themselves and their members apart from the rest of the university.

Since the most effective way to legitimate a new practice is to reward it, the reward structure of an organization is a good place to find reinforcement for its most important values. For example, ratings of faculty research performance or scholarship often take account of research funding obtained as a way to encourage that activity. One college of Hilltop Tech uses a formula that converts research dollars received by a faculty member into a rating of research performance. Departments have been struggling over ways to evaluate industrial support alongside more traditional federal grants and have had special difficulty deciding what to think of industrial support obtained by center administrators to support the research of a junior faculty member. For their part, center administrators have had to take the part of affiliated junior faculty in tenure cases, arguing the scholarly value of their center participation to skeptical department chairs.

The concept of the technocratic organization offers an accurate analytic description of UIRRs. Viewing centers in this framework makes their origins and growth part of larger trends in the political economy, not the opportunistic behavior of organizational leaders, nor the mystical workings of institutional forces. By connecting the rise of such organizations to the political economy, Heydebrand (1989) offers a rationale for their existence and an account of their origins. There is an appealing paradox at the core of this conception, a paradox or condition that some might call vaguely "postmodern." The idea is that organizations adapt to change by creating changeable parts of themselves, allowing them simultaneously to change and to remain as they were, to serve new ends while remaining loyal to existing purposes, to become different while remaining themselves. This shifts the attention of organizational research from the diversity or similarity of a population of organizations to the segmentation and internal differentiation of organizations taken one at a time. One then would look across organizations and over time for similar patterns of differentiation, arising from analytically similar circumstances.

A major shortcoming of the work presented here, both Heydebrand's ideas and the empirical study at Hilltop Tech, is that neither is very dynamic. We have a rich description of conditions and events, and we can fit that description into analytic categories, but we cannot say clearly where events lead. I address this somewhat speculatively in the conclusions. A second shortcoming of the idea of technocratic organizations is its connection to technology, computer technology in particular. I simply set aside that part of Heydebrand's exposition but return to it here to make the point that organizational structures are social technologies in themselves and need not be catalyzed or enabled by "hard" technologies. While computer integration or mediation might be consonant with the character and purposes of a technocratic organization, very often all the same organizational properties may be achieved without the computer.

## CONCLUSION

UIRRs may be considered technocratic organizations, that is, informal, permeable, flexible, and culturally controlled organizations that are agile, open, and responsive, they are the parts that change first and most often, while other parts remain constant or change slowly. For the most part, matters stand at a descriptive level, however, without a causal explanation or dynamic. In contrast, both resource dependence and institutional theory offer causal propositions. I offer the ideas below to remedy this.

Technocratic organizations offer a way around the forced choice of market control versus hierarchical control, a property they share with network organizations (Nohria and Eccles 1992; Powell Koput and Smith-Doerr 1996). Replacing, supplementing, or perhaps coexisting with control exercised through markets or hierarchies is another sort of control imposed by a web of relationships and commitments, a culture of process and performance. In the specific instance of UIRRs, such control is exerted through a web of contractual relationships to do projects, a network of promises to show results to corporate affiliates, and the set of reciprocal obligations of employment (for center staff and visiting engineers) or employment prospects (for students). A canopy of academic values, norms, and traditions contributes to a sense of solidarity and commonality of purpose. Reciprocal obligations of material exchange (persons, resources, problems, solutions) and of mutual obligation (to perform, to mentor, to collaborate) replace market discipline and imperative coordination.

The obligations of exchange echo, on a smaller scale, the core processes of resource dependency, while the matrix of norms and values seems an expression of institutional influences. More concretely again, in exchange for resources provided, industrial sponsors expect and usually receive solutions to problems and students with skills and interests that suit their needs. To do this, the UIRR—this technocratic organization—tilts its work and work process as needed. At a cultural level, industrial sponsors expect businesslike practices to be represented within the UIRRs, and they are (for example, in CBT's project teams). Yet they also expect academic values to be expressed and honored through education, graduation, participation, communication, publication, and the like. The academic value of open communication is not honored in word alone but is embraced and seen to have worth. For example, UIRRs offer a safe haven for competitors to share information, ideas, and problems that they are unable to share in other venues.

In closing, a few words about the dynamics of UIRRs, derived (admittedly, in freehand) from the model of the technocratic organization. In the first place, UIRRs undermine both bureaucratic and professional authority, implying that they will challenge the ability of both faculty and administrators to assert control. In the limit, of course, faculty may refuse to participate in centers, and administrators may close them. But short of such draconian measures, active centers will undermine traditional means of control. In the second place, UIRRs will erode boundaries within the university and between the university and its environment, creating a more integrative and collaborative working environment. Some traditional departments will be challenged, receiving a smaller relative share of resources or closing outright, while new programs and organizational subunits will grow up between or alongside them. Similarly, university boundaries will erode further, blurring the edges that once separated students from employees, members from outsiders, teaching from research, gifts from contracts, the university itself from those organizations in its network. In the third place, change will become the usual order of business for universities, but it will be a segmented change that does not necessarily involve all parts of the organization. Some parts will not change, perhaps by successful resistance, perhaps by design, perhaps through oversight or neglect. Finally, one must look beneath the appearance of stability for the evidence of change. At the surface it may appear that nothing is new, faculty will still work in departments, teach courses, conduct research, and write books and articles. Students will take courses, assist faculty in their research, and write term papers and theses. But a look beneath this tranquil surface will reveal change in the character of teaching and learning and writing, in the nature of interactions between students and faculty, and in the motivations, purposes and meanings of what they accomplish.

## NOTES

*Organizational Perspectives on University-Industry Research Relations* was prepared for the annual meetings of the Society for Social Studies of Science, November 1989. A prior version was presented at the annual meetings of the American Sociological Association. The research was supported by NSF BBS87-11341.

Mike Gunderloy gathered much of the historical and interview material reported here. I am very grateful to him for his excellent work, and for our many lengthy and insightful discussions.

1. Closely related to the resource dependence perspective is organizational ecology, which argues that the wide variety of organizational forms observed in contemporary society arose through the increasing complexity of environments (Hannan and Freeman, 1977). The chief difference between the two perspectives is that resource dependence theory predicts structural change in organizations, whereas organizational ecology proposes that aggregate change occurs through the creation and demise of organizations, not their transformation. Furthermore, organizational ecology predicts heterogeneity in populations of organizations and appeals to diversity and selection— the births and deaths of organizations—as a means of shaping organizational populations. Yet universities strongly resemble one another, often deliberately patterning themselves on successful programs elsewhere, and universities are rarely born and seldom die. (The striking exception to this is the strong presence in the mid-1990s of for-profit universities, such as the University of Phoenix.) Since the founding or demise of a university, particularly a university with any significant research activity, is a rare event, this pattern of explanation is not especially powerful.

2. The related business activities of nonprofit organizations are a case in point. Such activities allow organizations to retain nonprofit status, and to preserve the special character of their missions as nonprofits, while entering the profit-making sector in a limited way to raise funds.

2

# New Arenas for University Competition: Accumulative Advantage in Academic Patenting

JASON OWEN-SMITH

## INTRODUCTION

This chapter examines changes in the stratification order that governs competition among American research universities. Increased patenting and technology transfer efforts by these nonprofit research institutions have effected a shift in the terms of competition between universities. Commercial concern with private science has fractured the cross-university stratification order that governed success in the traditional activities of academic science, grantsmanship, and publication. Patenting and commercial success allowed some institutions to compete more effectively than was possible under the status order that governed post-cold war science. Nevertheless, I suggest that the stratification systems that govern commercial (private) and academic (public) science success are becoming more closely intertwined making it more difficult for Research One universities to succeed in either the private or the public realm alone. Enabled by key federal policy changes, transformations in the character of inter-university competition will require successful universities to navigate between public and private science instead of focusing solely on academic or commercial efforts.

In the last two decades American research universities have developed increasingly close ties to industry and commerce. Through technology transfer initiatives, strategic alliances, spin-off firms, venture capital activities, and intellectual property development, universities are becoming a driving force behind the development of regional economies (Feldman 1994; Feldman and Florida 1994; Jaffe, Trajtenberg and Henderson 1993) and high technology industries (Saxenian 1994; Rosengrant and Lampe 1992). Increasingly close ties between universities and firms have blurred accepted boundaries between science and technology. With this blurring, universities' traditional roles as producers of trained personnel and upstream, or basic, research are also subject to change (Powell and Owen-Smith 1998).

Recent research into the university's development questions the strict division of labor between universities and industry that has held credence since the end of World War II. Geiger (1986) argues that the "knowledge-plus" focus of American universities has been a reality for much of the twentieth century. Ben-David (1977) also notes that since their inception U.S. universities have had a more practical orientation than universities in Britain or Germany. Rosenberg and Nelson (1994, 190) argue that two major structural transitions have shaped the development of U.S. research universities.

The first shift began with the 1862 Morrill Act which established land grants for public institutions. Through the late nineteenth and early twentieth centuries, this change resulted in the "rise and institutionalization of the engineering disciplines and applied sciences as accepted areas of academic training and research" (Rosenberg and Nelson 1994, 213). The legitimization of research and training in industrially related areas "regularized" university-industry connections that had previously proceeded on an ad hoc, university-by-university basis.

The second transformation occurred after World War II with the rise of the "Cold-War/Health-War" policy coalition (Slaughter and Rhoades 1996). The coalition developed around nuclei in the "military-industrial complex" and national health care, resulting in massive federal funding increases for academic research (Slaughter and Rhoades 1996, 308). Rosenberg and Nelson (1994, 213) highlight dual effects of the postwar transformation. First, the federal funding influx shifted universities' research foci from local industrial needs to national concerns with defense and health. Secondly, academic science became more basic under the linear model of innovation that legitimized a division of labor between "upstream" university research and "downstream" industrial development.

I suggest that centralizing academic research funding in federal agencies had a third consequence. It enabled the development of a stable national stratification system among American research universities. The grant-based and peer reviewed funding architecture instituted under the cold war/health war regime may have created a system where a few elite institutions won the majority of federal Research and Development (R&D) grants.

Slaughter and Rhoades (1996) identify a third transition whose outcomes include changed relationships between industry and academia. Post-cold war economic competition shifted federal science policy from defense and health emphases to "competitiveness." The Cold-War/Health-War coalition aimed to "win the cold war and the war on disease," while the competitiveness regime aims to "[w]in global control of markets through privatization and commodification of science-based intellectual property" (Slaughter and Rhoades 1996, 308). Developing market control through intellectual property requires commercialization of outputs from America's premier research institutions, Research One (R1) universities.

The competitiveness rationale underlies universities' multiplex connections to the private sector, changing faculty roles, a blurred division of labor between academia and industry, and a new institutional role as "creators and retailers of intellectual property" (Chubin 1994, 126). The institution's transformation from a cradle of basic science to an engine of economic development (Feller 1990) is most obvious in attempts to privatize university science by patenting. The 1980 Bayh-Dole act (PL 96-517) allowed nonprofit organizations, including universities, to secure intellectual property (IP) rights to findings derived from federally funded research. Bayh-Dole encouraged "academic capitalism" (Slaughter and Leslie 1997) by catalyzing privatization through patenting.[1]

Expanded university IP activities indicate greater privatization of scientific results. The competitiveness rationale also changes scientific selection procedures by altering the grant review process (Cohen and Noll 1994; Powell and Owen-Smith 1998). Research results are privatized by patents, but selection reflects private science as granting agencies increase concern with proposals' technological and economic implications. The changing selection process suggests a shift in the nature of interuniversity competition for R&D funding. This chapter examines patenting's implications for that competition and for the institutional stratification system that came to structure research selection under the cold war regime.

An important consequence of the increasing university-industry linkages has gone unnoticed. Private science's implications have been examined for scientific careers (Blumenthal 1992; Blumenthal et al. 1996; Hackett [Science as Vocation] this volume), for the institutional mandate of universities (Feller 1990; Chubin 1994; Powell and Owen-Smith 1998), for the development of new organizational forms (Hackett [Organizational Perspectives, this volume]; Rhoades and Slaughter 1991), for university laboratories (Kleinman 1998), for the culture of science (Packer and Webster 1996), and for industrial and economic development (Mansfield 1991; Henderson, Jaffe, and Trajenberg 1998; Nelson 1996). Nevertheless, little attention has been paid to patenting's effects on competitive relationships among universities. I examine the relationship between stratification in private (patent based) and public (publication/reputation based) university competition with data combining patent information and measures of federal and industrial R&D funding for eighty-seven R1 universities.

## PUBLIC SCIENCE, PRIVATE SCIENCE

Several scholars emphasize the differences between patenting and publishing (Dasgupta and David 1987, 1994; Packer and Webster 1996; Rip 1986; Myers 1995). Patents are legal documents that guarantee an inventor limited monopoly rights to a new technology in exchange for full disclosure of the invention's details. To be patented, an innovation must meet three criteria: it must be novel, useful, and nonobvious. The latter criteria means that inventions are not patentable if patent examiners deem them to be obvious to a person skilled in the relevant art. In biotechnology, for instance, a person skilled in the art might hold an advanced degree in a related scientific discipline and work in the field in which the invention was made.

Patents convey exclusive rights to an invention for a period of seventeen years.[2] They do not, however, convey the right to make or use an invention. Instead patent rights allow owners to prevent others from using their invention. Patents are "fences of interest" (Rip 1986) that delimit an inventor's exclusive property rights. Unlike scientific publications, which enroll diverse groups in the collective development of public knowledge (Latour and Woolgar 1979; Callon 1986), patents advertise private knowledge whose development and legal use requires owner consent (Dasgupta and David 1987). Pecuniary rewards accrue to inventors whose patents demarcate valuable "plots" of knowledge. Like deeds to land, patents mark property lines. They are the primary mechanism for disseminating private science. Publications, in contrast, are the primary coin of the realm for public science.

Scientific articles are "funnels of interest" (Rip 1986) designed to encourage readers to pursue research in an area (Callon 1986). Researchers write articles for expert audiences engaged in work similar to their own. When articles are successful, skeptical readers take novel findings into account in their own work and reputational rewards accrue to the author (Arrow 1987; Stephan 1996; Chubin and Hackett 1990; Merton 1957, 1988). Scholarly publication not only ensures that new knowledge enters the public realm but also actively involves other researchers in its development.

The outputs of public and private research have different implications. The processes for securing patents and publications also differ. Packer and Webster (1996) highlight the differences between patenting and publishing scientific results. Where publications are tailored to a specialist peer audience, patents must be written for an educated layman, the patent examiner, and defended against the skepticism of a legal fiction, the "person of ordinary skill in the art." Standards of publication for patents and publications are very different. Myers (1995) quotes a frustrated scientist/inventor who concretizes some key differences between patenting and publishing. The scientist describes the rhetorical styles of papers and patents.

> . . . I make the argument of a scientific paper, e.g. "everybody knows about his problem, this is the prior work done, everybody knew this was the experiment we needed to do, and everybody knows we should expect such and such a result. Well I have now done that experiment and I got the result everybody expected and I did it very carefully and everybody's theory is correct."
> . . . whereas in a patent , the argument is "There was a need, but no one knew where to go and nothing was done that was very relevant. And out of the blue, by surprise"—the key thing is surprise—"exploring, we didn't know where we were going, and we stumbled on this completely surprising thing, and we realized its importance and we created some industrial use out of it." (85)

Myer's analysis (1995) focuses on the rhetorical differences between patents and publications, but in doing so he highlights several key differences between public and private science. The audience, rhetoric, and implications of research differ across the public and private science regimes. This difference exists despite the fact that patent applications and scientific articles may describe exactly the same set of findings. The variation in rhetoric supports Dasgupta and David's argument (1987) that the difference between the realms of public and private science have more to do with researcher's commitments to the norms of patenting or publishing than with actual differences in scientific practice.

Until recently, the institutional realms of private and public science were also distinct with publications remaining the primary territory of academic researchers while patents were concentrated among scientists in industry. The passage of Bayh-Dole and the development of the competitiveness funding rationale blurred both the institutional and rhetorical boundaries characteristic of public and private science (Powell and Owen-Smith 1998). But blurred distinctions do not completely mask the widely disparate process and implications for patenting and publishing. Increased overlap between realms simply implies that a set of players (universities), dominant on the field of public science, has begun to play in a new arena. But does changing the field and rules of the game also change the ranking of the players? If public and private science are closely linked, and success in one translates easily into success in another, then the answer is no. If, on the other hand, important cross-field differences exist, then opening a new competitive arena will have implications for institutional success and for the stable system of ranking and prestige that undergirds interuniversity competition for public science funding.

## STRATIFICATION IN SCIENCE FUNDING

Universities compete with one another for students, faculty, and especially for the prestige that accompanies publications and federal grants. That competition is structured by a stable stratification order. Success depends upon institutional reputation. Universities that have already reaped reputational and fiscal rewards are better positioned to succeed in future competitions because of their established reputations. Institutional chances for success are stratified by this accumulative advantage process. In 1968 R. K. Merton coined a term to describe the process. He called it the Matthew Effect from the gospel of Saint Matthew: "To those who have more shall be given."

Merton's concern was with the Matthew Effect's functionality for scientific reward and communication systems. Nevertheless, numerous studies support accumulative advantage stratification models. These models hold at the individual (Allison and Stewart 1974; Dey, Milem, and Berger 1997; Levin and Stephan 1991, 1998), institutional (De Solla Price 1963; Chubin and Hackett 1990; Bentley and Blackburn 1990; Schultz 1996), and even national levels (Bonitz, Brucker, and Scharnhorst 1997).

An extensive study of scientific stratification in the Mertonian tradition concluded, with Merton, that accumulative advantage exists in science but that it is functional rather than vicious (Cole and Cole 1973). Zuckerman's (1977) detailed interviews with American Nobel Laureates indicate that elite scientists are aware of accumulative advantage's effects and that they make efforts to manipulate the system in favor of themselves and their students (see also Merton 1968). Individual, institutional, and national chances for success in the competitive world of public science are stratified through accumulative advantage. In essence, the Matthew Effect argues that success follows success. The best predictor of who will succeed in competitions between universities is who has succeeded in the past.

Federal research funds are disproportionately disbursed to R1 universities. Eighty percent of federal R&D dollars goes to just 100 institutions (Schultz 1996; Chubin and Hackett 1990). Awards are also unevenly distributed within this group. Ten institutions receive 21 percent of federal grant dollars (Schultz 1996, 133). It should not be surprising that these elite institutions include the most successful cold-war and health-war universities. I conceptualize institutional stratification in federal funding competitions as an outcome of the post war centralization of science and engineering funding in a handful of federal agencies. Thus, the Matthew Effect holds for federal grants not only because a few agencies disburse the bulk of funds (Chubin and Hackett 1990; Slaughter and Rhoades 1996), but also because the grant review process is reputationally based.

Reviewers are aware of individual scientists' track records and of their institution's reputation. In this way universities with strong histories of grant success are accorded higher status than those without such histories. Success at obtaining federal funding breeds more success because of the reputational rewards that accrue to investigators and institutions who obtain significant grant support.[3] If institutional stratification in federal science and engineering (S&E) funding is a consequence of the cold-war/health-war science policy rationale, then the shift to a competitiveness rationale should have some effect on universities' chances at obtaining funding.

Resource dependency arguments have been offered to explain increasing university commercialization (see Hackett, chapter 1). From this perspective, policy changes are themselves an outcome of academic S&E's shrinking resource base. Fewer federal resources drove universities to seek other avenues for support. Hackett (chapter 1) concentrates his attention on industrial support for university S&E. He argues that there is a strong pattern of potential dependence whereby the fraction of support a university receives from industry varies inversely with the institution's level of federal support. Note, however, that this pattern of resource dependence is a consequence of an accumulative advantage model of stratification. Institutions who already have substantial federal funding are not driven to search for other sources, because they continue to receive high levels of federal support. Institutional stratification here ensures that the pinch of a changing resource base is most strongly felt at the bottom of the status order.

Where the resource dependency perspective focuses on structural relationships between organizations and their environments, neo-institutional theory (Dimaggio and Powell 1983) focuses more on symbolic relationships of legitimacy (Hackett, chapter 1). Under this perspective the shift to a competitiveness rationale is less a result of lobbying on the part of universities that suffer from resource scarcity than it is an indicator of a new symbolic mandate for the institution. Powell and Owen-Smith (1998) argue that the shift in university mandate from public to private science has consequences for the funding of university research. A commercially focused research selection process moves universities away from traditional peer-reviewed funding sources.

In addition to increasing industrial support, Powell and Owen-Smith (1998) trace a rise in reliance on federal "earmarks" for R&D that bypass peer review. Here too it is universities near the bottom of the institutional stratification order that turn to alternate funding sources. In a congressional hearing on federal earmarking, Jonathan Silber, past president of Boston University, defends his institution's aggressive pursuit of earmarked funds in terms that highlight the competitive outcomes of accumulative advantage in federal funding. Silber argues that earmarking represents a way to "level the playing field," charging that traditional peer review represents "a tight-knit old boy's network" and an "oligopoly." Silber considers earmarked funds to be ". . . the ante that gives places like Boston University a seat at the table, and the last thing those (elite) institutions want is an aggressive new player in their game" (U.S. Congress, House of Representatives 1993, 231). The imagery here is decidedly one of frustration with competition under a stratification order that differentially benefits a small group of elite institutions.

Whether we use an institutional or a resource dependence perspective to explain the increasing privatization of science in American research universities one outcome remains the same. The shift to a competitiveness rationale resulted in a wider array of funding sources for university research. These resources often bypass traditional peer-reviewed avenues and are most aggressively pursued by universities nearer to the bottom of the stratification order established in the postwar period of "big government" funded science (de Solla Price 1963). For whatever reason, a new playing field has opened for universities. Heralded by the passage of Bayh-Dole, IP and academic capitalism have changed the terms of competition between universities. What remains to be seen is whether the new competitive arena affects changes in the traditional stratification order.

## MEASURING SUCCESS IN PUBLIC AND PRIVATE SCIENCE

The Matthew Effect holds for public science through the mechanism of reputationally based peer review, but a strong consequence of universities' new institutional mandate is a move away from peer-reviewed funding sources. Industrial and earmarked support bypasses traditional peer review. Income derived from the successful IP development is also independent of reputation and peer review because the patenting process differs fundamentally from peer review (Packer and Webster 1996; Etzkowitz and Webster 1995). University royalties from IP licensing have grown exponentially in the years since Bayh-Dole. In 1997 academic licensing revenue reached $698.5 million, an increase of nearly 20 percent over patenting income in 1996 (Association of University Technology Managers, 1997, 2). This new and increasingly important source of revenue depends upon private science success. Only universities that successfully patent their research outcomes are positioned to take advantage of this new income stream.

Levels of federal S&E support are common and relatively unequivocal measures of success in public science. Patent volume, the number of patents assigned to an institution, is a common measure of inventive productivity for private science (Albert et al. 1991; Narin and Brietzman 1995; Schmookler 1966; Griliches 1990; Trajtenberg 1990; Pakes and Griliches 1980).[4]

Much attention has been directed toward university patenting in recent years. Henderson and colleagues (1998) compare characteristics of university patents with those of patents held by other institutions, finding that academic patents have become more like industrial patents over time, with only a small percentage offering significant economic returns. Blumenthal and colleagues (Blumenthal et al. 1986; Louis et al. 1989; Blumenthal 1992; Blumenthal et al. 1996) have extensively analyzed surveys of life sciences faculty finding that successful academic patentors are also highly likely to succeed in federal grant competitions. The United States Patent and Trademark Office (USPTO) released a technology assessment and forecast report in 1997 dedicated entirely to the analysis of academic utility patents. To reiterate my opening remarks, this chapter uses patent volume as a measure of university success at private science to expand those studies by: (1) focusing on a specific and highly productive segment of academia, R1 universities; (2) by examining patenting as an indicator of success in a new arena of interuniversity competition; and (3) by considering the relationship between public and private stratification orders.

Similar to success at obtaining federal funding, patenting success appears to be highly stratified. Since the passage of Bayh–Dole, American universities have witnessed an explosion of commercial activity. Academic patents increased by nearly 700 percent following Bayh–Dole, while total U.S. patenting grew by only 50 percent. By 1997 academic patents represented nearly 5 percent of all patents assigned to American nongovernmental organizations (USPTO 1997, 4). The lion's share of these patents are assigned to the country's most prolific research institutions, R1 universities. Even within this elite group, patenting success is unevenly distributed.

Figure 2.1 traces trends in the volume of patents assigned to eighty-seven U.S. R1 universities from 1976 to 1996. Note the steady increase in the mean number of patents assigned to all universities over this time period. More interestingly, note the higher starting point and steeper curve for the ten most prolific patenting institutions. At first glance, it appears that the Matthew Effect also holds for patent volume. Note also the explosion of patenting by the top ten in the late 1980s. Figure 2.1 implies that successful patentors remain successful. The gap between more and less successful institutions also increases significantly through the 1990s.

I suggest that the increase in this later period results from the development of a stable stratification system similar to that found for federal funding. Once such a stratification system is in place, the Matthew Effect ensures that those universities who have been successful will continue to be so. I argue that through the early 1980s the new field of competition, patenting, and private science was relatively open but that early success in this time period resulted in the development of an elite group of patenting universities. Once this group stabilized, accumulative advantage processes ensured their continued dominance, and the gap between "have" and "have-not" universities increased precipitously.

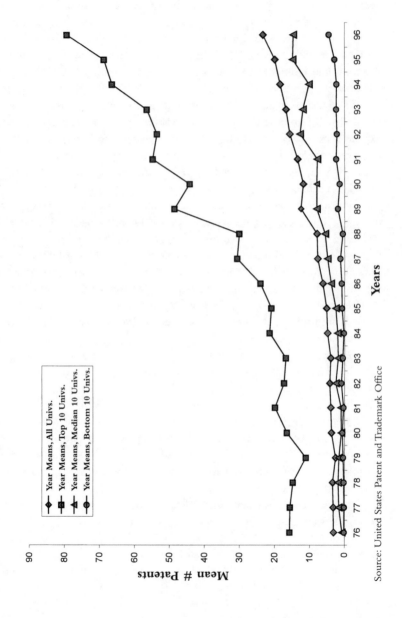

Figure 2.1. Mean Patents Assigned by Year: 1976–1996

Mean # Patents

Years

- ◆ Year Means, All Univs.
- ▪ Year Means, Top 10 Univs.
- ▲ Year Means, Median 10 Univs.
- ◉ Year Means, Bottom 10 Univs.

Source: United States Patent and Trademark Office

Figure 2.2 provides some descriptive support for this claim. It presents the between-university variance in patent volume from 1976 to 1996. Note the explosive growth in variance after 1989. The curve suggests that there has always been a patenting gap between universities. It also indicates that that gap has grown over time and that it widened explosively beginning in the 1990s. In the remainder of this paper I take up several questions related to the development of a new stratification order for the realm of private science: (1) Who are the private science "success stories?" (2) To what extent does success in the realm of public science translate into the realm of private science? (3) What factors explain success at patenting? (4) Do the same factors explain patenting success in both the 1980s and the 1990s? and (5) Do early (1980s) characteristics of universities and their patents explain later (1990s) patenting volume?

## INDICATORS AND DATA

My primary concerns in this chapter are to examine the development of a new arena for interuniversity competition, that is, private science, and to consider the extent to which success in public science translates into achievement in privatization. To that end I develop a data-set combining indicators of public science success with indicators of successful privatization for eighty-seven R1 universities.[5] Data on all variables were collected yearly for the periods from 1984 to 1987 and from 1994 to 1997.

### Public Science Indicators

I gathered data on three variables that serve as indicators of success in the arena of public science. All three were extracted from WebCASPAR, the Web-based data system maintained by the National Science Foundation (NSF). WebCASPAR is designed to make a wide range of institutional-level data regarding American higher education, science, and engineering available to researchers. The three indicators of public science achievement I chose are: (1) federal obligations for research and development, (2) federal obligations for training, and (3) the National Research Council's rating of the scholarly quality of a university's faculty for all science and engineering disciplines.

Federal obligations for R&D are the promised amounts, reported in thousands of dollars, of federal support for research and development. Values include all direct, indirect, incidental and related costs resulting from or necessary to R&D. This variable does not represent actual expenditures on R&D at universities. Instead, it reflects the amount of support promised to an institution in a given year. Because a key mechanism for the Matthew Effect is reputation, federal obligations are a better indicator of success than actual expenditures because

34

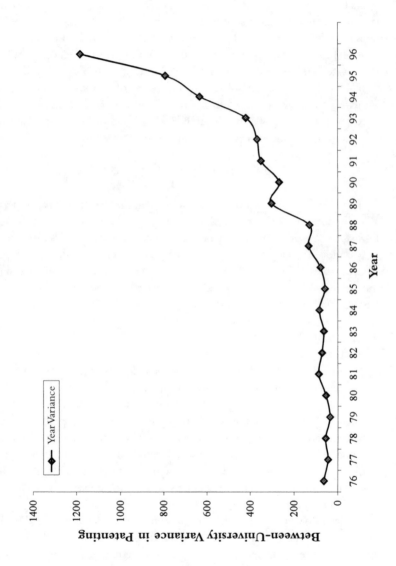

Figure 2.2 Between–University Variance in Patenting by Year: 1976–1996

obligations represent the level of support offered by federal agencies rather than the amount of federal funds that were actually used. Thus federal obligations are an indicator of an institution's relative reputation with federal granting agencies and reviewers.

Likewise, federal obligations for training do not represent actual expenditures. Instead, this variable captures the amount of federal support offered to an institution for fellowships, traineeships, and training grant programs that are directed to the development of scientific and technical manpower. This variable captures the level of support federal agencies offer to an institution for its graduate students and post-docs. It also proxies for the perceived quality of graduate and post-doctoral students and training because this variable is also reputationally based.

I use a third reputational variable as an indicator of the quality of an institution's science and engineering faculty. Approximately every five years the National Research Council (NRC) surveys scientists about program quality at American research institutions. The NRC's Scholarly Quality of Program Faculty rating indicates the general disciplinary esteem for a university's faculty. The NRC survey asks reviewers to rank program quality on two dimensions, the scholarly quality of a program's faculty and a program's effectiveness at educating research scientists. Ratings were made on a scale ranging from 0, "not sufficient for doctoral education," to 5, "distinguished." The aggregate measure for each university represents the mean rating of all science and engineering programs surveyed. For analyses of 1980s data, I use the NRC survey conducted in 1982 and reported in 1985. For 1990s analyses, I use the survey conducted in 1993 and reported in 1995.

## Private Science Indicators

I use two sets of indicators of private science success. The first set is composed of patent volume measures and industry R&D expenditures. The former were collected from the USPTO's bibliographic patent database, and the latter come from the WebCASPAR system. The second set reflects the characteristics of a university's patent portfolio and was developed by coding a complete set of university-assigned patents for the years 1985 and 1995.

Patent volume is a straightforward measure of the number of patents assigned to an institution or to its representative.[6] Under U.S. patent law only individual inventors can be awarded patents. Assignment of ownership is the mechanism by which organizations establish IP rights associated with patents awarded to their employees. The second indicator of private science success is a variable that indicates the value (in thousands of dollars) of all an institution's grants and contracts for R&D from profit-making institutions. This variable captures the extent to which R&D at a given institution is funded by industrial rather than federal sources.

My second set of indicators captures characteristics of an institution's patents. Data were coded from all patents assigned to R1 universities in 1985 (373 patents) and 1995 (1506 patents). For this analysis, five variables were chosen to capture the "quality" of patents and the organizational location of patenting in an institution. The first variable, number of foundation-assigned patents, measures the number of patents assigned to an associated foundation, for example, the University of Wisconsin's Wisconsin Alumni Research Foundation (WARF). This variable serves as an indicator of the extent to which expertise at navigating the patenting process, developed primarily through patenting experience, is internal or external to a university.

The second variable, number of patents jointly assigned with industry, measures the number of patents an institution holds that are jointly owned with a firm. A patent that is jointly assigned is jointly owned. Like co-authorship on a scientific paper, joint assignment indicates high-level collaboration and often, because patents represent legal ownership, represents an explicitly contractual agreement. While this variable serves as an important indicator of close collaboration with industry, it does not capture the wide range of variation in types of collaboration. Thus, a notable increase in the number of patents jointly assigned to industry would be notable precisely because joint assignment represents an extremely high level of collaboration and thus underestimates the actual amount of contact between universities and firms.

I also include a variable, mean prosecution time, that measures (in months) the mean time between an institution's filing of a patent application and issuing of a patent from that application. This variable indicates the average complexity and degree of controversy associated with a university's patent applications. More complex patents in less established technological areas take longer to complete the patent process than patents in well understood technological areas (Henderson et al. 1998; Narin 1994).

In addition to prosecution time, two other variables capture, albeit imperfectly, some aspects of patent quality. First I include a variable that measures the mean number of claims associated with a university's patents. Claims are the formal, one-sentence statements in every patent that legally establish the breadth of property rights afforded to that patent. In most cases the bulk of conflict in the prosecution process surrounds the claims section (Myers 1995) because the contents of this section establish the actual breadth of protection. Broader patents are more valuable than narrower patents because they convey exclusive rights to a wider array of applications and aspects of a technology. A variable measuring number of claims serves as a rough proxy for patent breadth. More claims are assumed to indicate patents conveying wider protection.

Finally I include a variable that measures the mean number of nonpatent citations in the prior art sections of a university's patents. *Prior art* is a term that includes all patents (foreign and domestic) and other sources that a patent examiner determines represent precursors to the innovation being patented. Prior art citations legally establish the novelty criteria for patenting, and they have been taken to indicate connections between patented innovations and inventors in much the same way as citations in scientific papers are used to indicate connections between findings and authors (Narin 1994; Podolny, Stuart, and Hannan 1996).

As the name suggests, nonpatent citations indicate the dependence of a patented innovation upon prior art published in nonpatent sources. While these citations can and do include any and every type of published material, among the most interesting are citations to scientific publications. Citation counts of this sort have been validated as measures of industrially important patents (Albert et al. 1990) and, perhaps more interestingly, have been used repeatedly to demonstrate the connections between public science and technology (Narin, and Olivastro 1992; Narin Hamilton, and Olivastro 1997; Narin and Olivastro 1998). By capturing the mean number of nonpatent citations in academic patents, this variable serves as a rough proxy for the closeness of fit between the more "basic" scientific research traditionally pursued by universities and the more "applied" research characteristic of patenting and private science.

Table 2.1 presents summary statistics for these indicators of university success in public and private science for 1985 and 1995. Several interesting features stand out on table 2.1. Note the changes from 1985 to 1995 in the patent characteristics variables. Over this decade the proportion of university patents assigned to foundations rather than directly to universities decreased from nearly 46 percent to about 22 percent. This decrease is indicative of the development of more in-house capacities to prosecute and manage intellectual property at R1 universities. The change suggests that offices of technology transfer and other such organizational units within universities are becoming more important for academic patenting than "buffer institutions" (Rosenberg and Nelson 1994) such as university foundations and research corporations.

At the same time the large increase in patents jointly assigned with industry, from less than 1 percent in 1985 to nearly 7 percent in 1995, indicates a greater amount of close collaboration between universities and firms in private science research. Interestingly, the lessening importance of buffer institutions and the increasing degree of direct university-industry collaboration are contemporaneous with indications of increasingly close ties between academic patents and scientific publications. The number of citations from patents to nonpatent sources has nearly tripled in the decade from 1985 to 1995. These descriptive statistics imply changes in the character of university patenting over the last decade.

TABLE 2.1

## TOP TEN PATENTORS RANKED ON TWO
## PUBLIC SCIENCE MEASURES, 1985 AND 1995

| University | Patent Rank | 1985 Grant Rank | NRC Rank[a] |
|---|---|---|---|
| MIT | 1 | 3 | 1 |
| Stanford University | 2 | 4 | 5 |
| Iowa State University | 2 | 81 | 48 |
| Cornell University | 4 | 9 | 14 |
| California Inst. Technology | 5 | 32 | 2 |
| Wisconsin - Madison | 6 | 8 | 13 |
| UC-Berkeley & UCSF[b] | 7 | 2 | 6 |
| Johns Hopkins | 8 | 1 | 31 |
| Georgia Tech | 9 | 33 | 41 |
| University of Minnesota | 9 | 15 | 19 |
| University of Utah | 9 | 38 | 56 |
| University of Texas - Austin | 9 | 21 | 21 |

| University | Patent Rank | 1995 Grant Rank | NRC Rank |
|---|---|---|---|
| UC-Berkeley & UCSF | 1 | 2 | 2 |
| MIT | 2 | 4 | 1 |
| Wisconsin - Madison | 3 | 9 | 14 |
| University of Florida | 3 | 43 | 54 |
| Stanford University | 5 | 5 | 5 |
| Cornell University | 6 | 13 | 9 |
| Iowa State University | 7 | 71 | 64 |
| California Inst. Technology | 8 | 29 | 3 |
| University of Texas - Austin | 9 | 27 | 17 |
| North Carolina State | 10 | 56 | 42 |

a: 1982 and 1993 Scholarly Quality of Faculty Rating
b: Rankings represent sums of patents and grants issued to UC-Berkeley and UCSF and the
mean of the two institutions' NRC rating.

## ANALYSIS AND METHODS

The shifting character of academic patents suggests growing links between public
and private science. Universities are becoming more directly involved in private
science, and academic patents are more closely tied to published findings, indicating closer links between the outputs of public and private science. Close col-

TABLE 2.2

## DESCRIPTIVE STATISTICS, PUBLIC AND PRIVATE SCIENCE INDICATORS, 1985 AND 1995

| | 1985 | | 1995 | |
|---|---|---|---|---|
| VARIABLE | MEAN | S.D. | MEAN | S.D. |
| *Public Science Indicators*[a] | | | | |
| Fed R&D Obs. | 58108.76 | 48687.52 | 106759.3 | 86928.46 |
| Fed Training Obs. | 2539.09 | 3026.77 | 5437.43 | 4999.11 |
| NRC Quality Rating | 3.044 | 0.761 | 3.164 | 0.632 |
| *Private Science Indicators* | | | | |
| Number of Patents | 4.287 | 6.024 | 5.402 | 7.321 |
| Number of Claims | 14.839 | 11.915 | 17 | 12.885 |
| Number of Non-Patent Citations | 3.876 | 6.641 | 10.502 | 14.269 |
| Prosecution Time[b] | 31.113 | 10.852 | 28.338 | 12.387 |
| | #(% TOTAL) | | #(% TOTAL) | |
| Number Foundation Assigned Pats. | 171 (45.8) | | 355 (22.2) | |
| Number Industry Assigned Pats. | 2 (.5) | | 104 (6.9) | |

a: All dollar amounts reported in thousands
b: Time in months from patent filing to patent issue

laborations between firms and universities are on the rise, as evidenced by increasing joint assignments. All of these descriptive changes support Powell's and Owen-Smith's (1998) contention that traditional distinctions between the realms of science and technology are breaking down. Changes in the characteristics of university patents do not, in themselves, however, indicate changes in interuniversity competition or in the institutional stratification order characteristic of public science.

Table 2.2 presents the ten (ties inclusive) most prolific patenting universities in 1985 and 1995 along with their rank on two key indicators of public science success: grant income and NRC quality ranking. This table indicates that achievement in the private science arena is not always matched by success in reputationally based arenas typical of public science. Note that for both 1985 and 1995 only about one half of the most prolific academic patentors are also among the top ten universities in grant income or quality rating. It appears that one set of private science "success stories" built significant patent portfolios while maintaining elite status in public science. These institutions remained among the most prolific patentors a decade later, indicating that an accumulative advantage process may be at work for patenting as well as for grants.

Other patenting institutions have developed significant IP portfolios while remaining nearer to the middle or even the bottom of the public science status order. Institutions such as Georgia Tech, Iowa State University, and the University of Texas at Austin demonstrate that the realm of private science presents opportunities for success to institutions that have not been spectacularly successful in the public (academic) realm. Table 2.2 also suggests that early success in patenting does not always guarantee continued achievement. In 1995 only two of the four institutions mentioned above remained among the top patentors. They were joined, however, by another pair of institutions who were distinguished patentors without being elite on public science criteria, the University of Florida and North Carolina State University.

The lesson of table 2.2 appears to be that the Matthew Effect holds, to some extent, both in and across private and public science. Nevertheless, the table does not support the contention that elite status in the realm of public science confers achievement in private science. It appears instead that the new realm of private science presents opportunities for success to universities whose low positions in the stratification system of public science would make it difficult for them to succeed in the traditional academic arena. For academic institutions, private science in the mid-1980s and 1990s appears to be a partially open field, suggesting that, while connected to public science, it is separate enough to allow more open competition between universities than is possible under the public science stratification system. To paraphrase Jonathan Silber, the realm of private science appears to make room for "new players at the table."

To further examine the relationships between the realms of public and private science, I turn to three sets of OLS regression analyses. The first set of analyses presents three nested models of patent volume at the university level of analysis in the mid-1980s. I regress indicators of public and private science success and control variables from 1985 on 1987 patent volume. Very similar models are presented for the mid-1990s, where explanatory variables from 1995 are regressed on 1997 patent volume. These models shed light upon changes in the relationship between public and private science from the 1980s to the 1990s. Finally, a model is run with a ten-year lag to determine the longevity of variable effects and the long-term relationship between success in public science and success in private science. This model regresses explanatory and control variables from 1985 on patent volume in 1995 at the university level.[7]

## 1980S MODELS

For the mid 1980s, I model university patent volume in 1987 using as independent variables the indicators of public and private success from 1985.[8] I include as controls demographic variables intended to capture the size of each institution's population of researchers under the assumption that more scientists create more potentially patentable findings. For this period, I present three nested models to allow for progressive tests of model fit. Model one includes only the nonpatent characteristic indicators of public and private science success. In all models, I log transform monetary values to capture the effects of percentage rather than unit changes.[9]

> Model 1:
> (Patent Volume)87 = b1(Patent Volume)85 + b2 (Log Fed. R&D Obl.)85 + b3 (Log Fed. Training Obl.)85 + b4(Log Industry R&D Exp.)85 + b5 (NRC Rating)82 + e

Model 2 builds on Model 1 by adding demographic control variables, including geographic region,[10] institutional control,[11] number of campuses,[12] and the number of science and engineering faculty, post-docs and graduate students such that:

> Model 2:
> (Patent Volume)87 = b1(Patent Volume)85 + b2 (Log Fed. R&D Obl.)85 + b3 (Log Fed. Training Obl.)85 + b4(Log(Industry R&D Exp.)85 + b5 (NRC Rating)82 + b6(# S&E Faculty)85 + b7(# S&E Post-docs)85 + b8(# S&E Grad Students)85 + b9 (East) + b10 (West) + b11 (Multicampus) + b12 (Private) + e

The full model, model 3, adds all but one of the patent characteristic variables to model two. I do not include the number of patents jointly assigned with industry because only two patents were jointly assigned in 1985. This variable will be included in models with two-year lags for the 1990s

> Model 3:
> (Patent Volume)87 = b1(Patent Volume)85 + b2 (Log Fed. R&D Obl.)85 + b3 (Log Fed. Training Obl.)85 + b4(Log Industry R&D Exp.)85 + b5 (NRC Rating)82 + b6 (# Foundation Pats.) + b7(# Claims) + b8(# Article Cites) + b9 (Prosecution time) + b10(# S&E Faculty)85 + b11(# S&E Post-docs)85 + b12(# S&E Grad Students)85 + b13 (East) + b14 (West) + b15 (Multicampus) + b16 (Private) + e

Model fit was tested for each model using progressive full and reduced model f-tests at the .05 significance level.

## HYPOTHESES

I present these regression models to shed some light on the relationship between the realms of public and private science in two time periods following the shift to a competitiveness rationale for S&E funding. I first present a number of coefficient-level hypotheses for the indicators. If opening a new realm to universities does affect changes in competition between universities, then much of the analytic action in these models will be found in inferences based on patterns of support for the coefficient-level hypotheses across the time periods examined.

If the Matthew Effect holds across the realms of public and private science and those who "have" in the traditional realm also "receive" in the new arena, then indicators of public science success should be positively related to later patent volume in both the 1980s and 1990s. Simply stated, the "cross-realm Matthew Effect hypotheses" predict that both log of federal obligations for R&D and NRC quality ratings will be positive and significant in all models. In terms of the regression models presented above: H1: $b2 > 0$, H2: $b5 > 0$. For these hypotheses, the null ($b1 = b2 = 0$) represents a case where the realms of public and private science are largely independent of one another.

If success in the realm of private science is independent of the established stratification order in public science and if accumulative advantage models of stratification hold for private science, then I expect indicators of private science achievement to be positive and significant in all models. More simply, if the Matthew Effect holds for patenting, then the best explanator of patenting success is earlier patenting success. To the extent that industrial R&D support is also indicative of private science success, it too should have a positive and significant coefficient in all models. In regression terms: H3: $b1 > 0$, $b4 > 0$.

Finally, I turn to hypotheses about the characteristics of university patents. I include patent characteristic variables in these models primarily to capture the effects of organizational learning and to indicate the value of a university's patent portfolio. Accumulative advantage in public science proceeds primarily through the mechanism of status and institutional reputation. However, the patent process is not dependent either on peer evaluations, reputation, or prior track record.

If the Matthew Effect holds for university patenting, then it must proceed through another mechanism. I suggest that organizational learning provides the mechanism for accumulative advantage in patenting. Universities that develop organizational competency in IP prosecution and management will be better able to identify and prosecute patentable innovations than universities that have not developed such competencies. Organizations learn by doing. Thus, hypothesis three, that early patenting success breeds later patenting success, captures this aspect of the Matthew Effect. Nevertheless, universities are not individuals. For an organization like a university to "learn" requires not only that it have experience, but also that it have some organizational mechanism for capturing and developing that experience. One such organizational mechanism might be an office of technology transfer (OTT). At this time direct data on the size, organization, and staff of Research 1 OTTs is not available to me.[13] Thus, I include a rough proxy in my models, the number of patents assigned to buffer organizations, such as foundations. By their very nature, university research foundations and corporations are separate from the universities themselves. This separation allows foundations to serve as buffers between public nonprofit institutions and private for-profit firms. One consequence of such buffering, however, is that it is the foundation rather than the university that develops competencies in identifying promising innovations and prosecuting complex patents. Thus I argue that the number of patents assigned to foundations should have a negative effect on patent volume. In regression terms: H5: $b6 < 0$.

Patents are valuable to the extent that they provide broad intellectual property protection for innovations that can be developed for sale or licensing. My best direct measure of a patent's breadth is the number of claims it contains. I assume, perhaps simplistically, that more claims equal more protection. If patent breadth is another measure of private science success and if the Matthew Effect holds for private science, then I expect a university's mean number of patent claims to be positively related to patent volume. Thus: H6: $b7 > 0$.

## FINDINGS

Tables 2.3 and 2.4 present the results of regression models with two- and ten-year lags respectively. Consider table 3. The first three columns of this table report the coefficients of three nested models of 1987 patent volume. Each model represents a significant increase in goodness of fit over the one that preceded it. Model three, the patent characteristics model, is best fitting. This model provides mixed support for my six hypotheses. The Matthew Effect does not appear to hold across the realms of public and private science in the mid 1980s. Neither hypothesis one, which postulates a positive relationship between federal R&D support and patent volume, nor hypothesis two, which proposes a positive relationship between faculty quality rating and patent volume, receive support from this model.

The NRC's faculty quality rating has no significant effect on patent volume, while log-transformed federal R&D obligations (grant income) have a negative effect. This latter finding suggests that soon after the Bayh-Dole act universities with less federal grant support patented more than universities with extensive grant support, lending some credence to the resource dependence argument for university privatization. Federal training obligations have a positive and significant effect on patent volume, suggesting that universities with higher levels of training funding are more accomplished at later patenting. This may be the case because universities with more, and more prestigious, training funding can attract more and better graduate students and post-docs. This

TABLE 2.3

## RESULTS OF OLS MODELS OF PATENTING WITH TWO-YEAR LAGS

| VAIABLES D.V. | MODEL1 # 1987 PATENTS | MODEL 2 # 1987 PATENTS | MODEL 3 # 1987 PATENTS | MODEL 4 # 1987 PATENTS | MODEL 5 # 1987 PATENTS | MODEL 6 # 1987 PATENTS |
|---|---|---|---|---|---|---|
| Intercept | 6.300 | 14.091 | 26.405* | -61.936** | -27.975 | -28.973 |
| S.E. | 11.323 | 13.756 | 12.732 | 19.845 | 22.734 | 24.680 |
| # of Patents[a] | 1.321*** | 1.313*** | 1.607*** | 1.107*** | 1.063*** | 1.065*** |
|  | 0.102 | 0.110 | 0.123 | 0.065 | 0.065 | 0.075 |
| Log of Fed. R&D Obl. | -2.332 | -2.942+ | -4.301** | 6.437* | 6.229* | 6.598* |
|  | 1.582 | 1.743 | 1.612 | 2.882 | 2.860 | 3.008 |
| Log of Fed. Training Obl. | 2.535** | 2.611** | 2.700** | 1.188 | -2.496 | -2.504 |
|  | 0.764 | 0.861 | 0.788 | 1.982 | 2.089 | 2.164 |
| Log of Industry R&D Exp. | 0.299 | 0.246 | 0.177 | -1.529 | -2.295+ | -2.328+ |
|  | 0.641 | 0.689 | 0.617 | 1.233 | 1.239 | 1.323 |
| NRC Quality Ranking | -0.302 | -0.451 | -0.169 | -0.637 | -1.647 | -1.883 |
|  | 1.103 | 1.312 | 1.261 | 2.455 | 2.441 | 2.595 |
| # Industry Assigned Pats. |  |  |  |  |  | -0.125 |
|  |  |  |  |  |  | 0.560 |
| # Foundation Assigned Pats. |  |  | -0.749*** |  |  | 0.099 |
|  |  |  | 0.173 |  |  | 0.119 |
| Mean # of Claims |  |  | -0.015 |  |  | -0.078 |
|  |  |  | 0.067 |  |  | 0.153 |
| Mean # Article Cites |  |  | 0.131 |  |  | -0.013 |
|  |  |  | 0.109 |  |  | 0.159 |

| VAIABLES<br>D.V. | MODEL1<br># 1987 PATENTS | MODEL 2<br># 1987 PATENTS | MODEL 3<br># 1987 PATENTS | MODEL 4<br># 1987 PATENTS | MODEL 5<br># 1987 PATENTS | MODEL 6<br># 1987 PATENTS |
|---|---|---|---|---|---|---|
| Mean Prosecution Time | | | -0.034 | | | -0.012 |
| | | | 0.043 | | | 0.186 |
| # of S&E Faculty | | -0.004* | -0.002 | | 0.005+ | 0.004 |
| | | 0.002 | 0.002 | | 0.002 | 0.003 |
| # of S&E Post-Docs | | 0.000 | 0.000 | | 0.015*** | 0.016*** |
| | | 0.004 | 0.004 | | 0.004 | 0.004 |
| £ of S&E Grad Students | | 0.001 | 0.001 | | -0.001 | -0.001 |
| | | 0.001 | 0.001 | | 0.001 | 0.002 |
| East Cost Region | | 2.520* | 2.053+ | | 0.577 | 0.694 |
| | | 1.301 | 1.185 | | 2.298 | 2.407 |
| West Cost Region | | 0.459 | -0.603 | | -0.031 | 0.158 |
| | | 1.366 | 1.263 | | 1.734 | 1.900 |
| Multicampus University | | 1.450 | 1.935 | | 2.272 | 2.461 |
| | | 1.406 | 1.289 | | 2.354 | 2.498 |
| Private University | | -0.386 | -0.213 | | 0.667 | 0.682 |
| | | 1.853 | 1.699 | | 2.7530 | 2.941 |
| $R^2$ | 0.7813 | 0.8218 | 0.8602 | 0.8805 | 0.9081 | 0.9097 |
| N | 73 | 73 | 73 | 81 | 81 | 81 |

a: in 1987 models all time varying variables are for 1985, variables are 1995 values for 1997 models.

+ p<.10, * p<.05, ** p<.001

TABLE 2.4

## OLS MODELS OF PATENTING VOLUME
## WITH TEN-YEAR LAGS

| VARIABLES D.V. | MODEL 1 # 1995 PATENTS | MODEL 2 # 1995 PATENTS | MODEL 3 # 1995 PATENTS |
|---|---|---|---|
| Intercept | -11.311 | -2.813 | 12.793 |
| S.E. | 20.546 | 24.418 | 24.141 |
| # of Patents[a] | 1.711*** | 1.806*** | 2.241*** |
|  | 0.185 | 0.195 | 0.264 |
| Log of Fe. R&D Obl. | 0.038 | -1.282 | -2.104 |
|  | 2.871 | 3.095 | 3.057 |
| Log of Fed. Training Exp. | 1.545 | 1.801 | 1.373 |
|  | 1.387 | 1.529 | 1.494 |
| Log of Industry R&D Exp. | 0.689 | 0.178 | -0.037 |
|  | 1.163 | 1.224 | 1.172 |
| NRC Quality Ranking | 1.123 | 2.188 | 1.198 |
|  | 2.002 | 2.329 | 2.391 |
| # of Foundation Assigned Pats |  |  | -0.869** |
|  |  |  | 0.328 |
| Mean # of Claims |  |  | -0.031 |
|  |  |  | 0.128 |
| Mean # Article Cites |  |  | -0.004 |
|  |  |  | 0.207 |
| Mean Prosecution Time |  |  | -0.123 |
|  |  |  | 0.081 |
| # of S&E Faculty |  | 0.003 | 0.004 |
|  |  | 0.003 | 0.003 |
| # of S&E Post-Docs |  | -0.001 | 0.002 |
|  |  | 0.007 | 0.007 |
| # of S&E Grad Students |  | 0.000 | 0.001 |
|  |  | 0.002 | 0.002 |
| East Coast Region |  | 30808 | 2.784 |
|  |  | 2.310 | 2.247 |
| West Coast Region |  | -2.182 | -3.375 |
|  |  | 2.426 | 2.394 |
| Multicampus University |  | 1.542 | 2.513 |
|  |  | 2.495 | 2.445 |
| Provate University |  | -0.263 | 0.079 |
|  |  | 3.290 | 3.222 |
| $R^2$ | 0.6854 | 0.7376 | 0.7764 |
| N | 73 | 73 | 73 |

a: All time varying independent variables are 1985 values
+ p<.10, *p<.05, **p<.01, ***p<.001

would especially be the case if younger scientists are more likely to pursue private science than established senior researchers.

Hypotheses three and four predict that two indicators of private science success, industry R&D expenditures and earlier patent volume, will have positive effects on later patenting success. The hypotheses also received mixed support in model three. The 1985 patent volume is clearly a robust predictor of 1987 patent success. At least within the realm of patenting, accumulative advantage is supported for the mid-1980s. The strong positive relationship between 1985 and 1987 patent volume suggests that successful privatization breeds continued success in the short term. This finding also tangentially supports my argument that the Matthew Effect in patenting proceeds through organizational learning rather than through reputation. Universities develop expertise at privatizing research results through patenting. Those institutions that have more successful patent applications can reasonably be expected to have developed greater competency at navigating the prosecution process than those who have not been as successful.

Hypothesis four predicts that as the amount of R&D funded by industry increases, patenting will increase. It receives no support from model three. In the mid-1980s it appears that university patenting success is separate from success in gaining industrial R&D support. This finding also implies that there are many different and potentially unrelated avenues to success in the realm of private science. If, as Hackett (chapter 2) suggests, resource pressures drive universities to seek industrial funding, then the null relationship between levels of industrial funding and patent volume implies that patenting and seeking industrial R&D support are distinct avenues of resource development. Either different pressures drive universities to patent or to seek industrial support, or institutions respond differently to the same resource pressures with some choosing to commercialize research through R&D support, while others privatize findings by patenting and pursue income through licensing.

Hypothesis five predicts that the number of patents a university assigns to buffer institutions, such as foundations, will be negatively related to patent volume. The hypothesis is supported in model three. This finding further buoys the argument that accumulative advantage in patenting, at least early on, proceeds through organizational learning mechanisms. Universities that patent extensively develop greater competency at identifying and patenting economically viable innovations, but institutions that accomplish this largely through "arms-length" mechanisms, such as buffer organizations, do not develop strong competencies in-house. Finally, hypothesis six, that mean patent breadth is positively related to patent volume, receives no support in model three.

Consider the last three columns of table 2.3. Models four through six report results of regressions on 1997 patent volume using 1995 indicators. Unlike models one through three, these models do not represent progressive increases in goodness of fit. Model 5, which includes controls but not patent characteristics variables, is the best-fitting model of the three. Including patent characteristic variables (in model six) does not contribute significantly to model fit. This suggests that the characteristics of university patents themselves have become less important over time. This finding is in line with Henderson and colleagues (Henderson et al. 1998) claim that university patents have over time become more similar both to each other and to nonacademic patents.

Consider Model five. Note first that neither of the patent characteristics hypotheses (H5 and H6) is supported by the model. However, the pattern of support for hypotheses one through four is very different from that found in similar models for the mid-1980s. Model five suggests that by the mid-1990s the realms of public and private science had become more closely related. More to the point, the model indicates that greater amounts of federal R&D support have a positive impact on patent volume in the mid-1990s.

This is a complete reversal from model three, which found that federal R&D obligations were negatively related to patent volume ten years earlier. The finding is suggestive of the changing relationship between the realms of public and private science. In the mid-1980s, soon after Bayh-Dole catalyzed the privatization of university research, the public and private science realms appeared to be separate and negatively related. Findings in the 1980s models buoyed the claim that private science represented a new competitive field for universities and underscored the argument that universities who successfully pursued research privatization through patenting did so because of difficulty in gaining sufficient support through federal sources.

The changed relationship between federal R&D obligations and patent volume in the 1990s suggests that by the later decade success in the realms of public and private science had become mutually supportive. Recall the steep increase in between-university patenting variance demonstrated in figure 2.2. After about 1990 the "patent gap" between more and less successful universities increased at a steep rate. This increase may be a result of increasing blurring between public and private science activities. Growing overlap between the realms allows for accumulative advantage to function not only within but across public and private arenas. This suggests a magnifying effect where successful grant getting breeds more federal support and increased privatization which, in turn, may lead to increased levels of grant support.

Hypothesis three, that prior patenting success is positively related to later patent volume, also finds strong support in the mid-1990s, lending further credence to the idea that the increasing gulf between "have" and "have not" universities is a result of accumulative advantage across overlapping public and private realms. Model five fails to support hypothesis five. In the mid-1990s industry

R&D support is negatively related to later patent volume, though the regression coefficient is only marginally significant ($p<.10$). While I will not make much hay over a weak relationship, the negative result implies that there may be multiple independent routes to university commercialization through industry support and through patenting.

Finally, consider table 2.4. This table reports similarly specified models across a decade's lag, regressing 1985 indicators on 1995 patent volume. As was the case for the mid-1980s, the models are progressively better fitting. A glance at model nine suggests that over the long term few indicators of either public or private science success are strongly related to patent volume. Only two of my original six hypotheses (H3 and H5) are supported across the longer time lag. The robust positive relationship between 1985 and 1995 patent volume indicates the strength of accumulative advantage in research privatization over the longer term. Combined with the negative relationship between reliance on buffer institutions (foundations) to prosecute and manage patents, this finding further advances the theory that organizational learning is the mechanism by which accumulative advantage progresses in the private science arena.

Organizational competencies are located in the standard operating procedures (SOPs) and structures of units, such as OTTs, whose focus is on the identification, prosecution, and management of intellectual property. Unlike federal or industrial research support which shows strong effects in the short term but not in the long term, both patents and the organizational expertise successful patenting represents do not change significantly over the course of a decade. Unless a university chooses not to renew them, patents issued in 1985 are still potential income gainers in 1995. More to the point, the SOPs and organizational units that house institutional knowledge about privatization are also long lived. Organizational expertise, developed through patenting, represents a form of intellectual capital that increases with use.[14] While the Matthew Effect in public science is largely reputationally based, the positive effect of patenting and the negative effect of outsourcing privatization activities over the longterm suggest that for private science accumulative advantage proceeds through organizational learning.

## CONCLUSION

The post-World War II period heralded massive structural change in the funding architecture for U.S. science and engineering. In addition to centralizing federal S and E funding, the Cold-War/Health-War policy rationale saw the rise of a distinct division of labor between universities and industry, with the academy conducting basic research disseminated through publications and industry focusing on applied research and patenting. I argue that this period also saw the development of a stable institutional stratification order for American research universities. Through reputationally based accumulative advantage processes, this system structured interuniversity competition for federal grant funding resulting in a game that differentially advantaged a few elite institutions.

The post-cold war era saw the development of a new science policy coalition. Driven by increased international economic competition and a shrinking resource base, the new "competitiveness" rationale emphasized privatization and commercialization of university research for economic gain. The 1980 Bayh-Dole act catalyzed academic privatization through patenting and heralded a new era of "economic capitalism" (Slaughter and Leslie 1997). In addition to altering the institutional mandate of research universities, the shift to a competitiveness policy rationale opened a new arena for interuniversity competition: private science. In the 1980s a new field, patenting, was opened to an old set of players, universities. Differences in emphasis, process, and rhetoric between public and private science ensured enough separation between the realms to allow competition only loosely by the reputational stratification order that constrained institutional chances for success in public science.

I argue that soon after the passage of Bayh-Dole the realms of public and private science were distinct enough to allow research universities that had not been spectacularly successful in public science competitions to develop expertise in the primary activity of private science, that is, patenting. The most prolific patenting institutions were comprised of both elite and nonelite public science universities. During the early stages of privatization, then, patenting represented a relatively open field free from the reputationally based constraints of public science competition.

A decade later, however, the realms of private and public science had become more closely intertwined. A private science stratification order governed by accumulative advantage processes developed through the mechanism of organizational learning. Preliminary evidence suggests that by the mid-1990s private science success was stratified in a manner similar to public science. Furthermore, by the 1990s, the increasingly blurred distinctions between science and technology, public and private, enabled convergence between the realms. As a result, the stratification orders characteristic of each arena became mutually supportive, amplifying the Matthew Effect and sparking an explosion in the patenting gap between universities.

My findings suggest that early development of in-house patenting expertise by some universities enabled the development of a patenting elite that overlapped but did not replicate the top of the public science stratification system. Opening the realm of private science to universities had the effect of restructuring competition among the institutions. The opening of a new realm to universities had the effect of offering a "seat at the table" to some nonelite institutions, but it may also have created a situation in which reputational and learning-based accumulative advantage are mutually supportive, making it increasingly difficult to attain excellence in either public or private science alone.

Treating public and private science as fields for interuniversity competition suggests a number of interesting directions for further research. First, aggregate-level statistical analyses of the sort presented here focus attention on system-wide outcomes while ignoring the historically contingent trajectories that led individual institutions to develop commercial competencies. Indeed, the most interesting cases of successful or attempted university commercialization may be subsumed in the error term in my regressions. This system-level focus also leaves open the important question of how differences in organizational arrangements for technology transfer affect commercial outcomes and cross-university stratification. I suggested that further analysis of licensing data and institutional data on the size, organization, and staffing of university offices of technology transfer could serve to deepen our understanding of how adminis-trative choices and historical accidents led some universities to commercialize successfully while others did not.

Focusing on organizational differences among universities should also serve to deepen analyses of accumulative advantage in private science. I argue that organizational learning differences enable accumulative advantage in patenting. For my argument to hold, actual differences in organizational learning expertise must exist. Conducting further processual examinations of technology transfer and patenting differently situated universities may shed light on the means by which research institutions learn to navigate the waters of private science. Finally, a strong consequence of this analysis is that honored distinctions between pub-lic and private science are breaking down. As these realms become more closely linked, it will become increasingly difficult for universities to succeed in either realm alone. This suggests that attaining and maintaining elite status in an age of competitiveness will require institutions to balance the demands and cultures of public and private science. There are undoubtedly many ways to integrate commercial and academic activities on university campuses. Further study of those strategies is necessary if we are to fully understand the institutional changes affecting competition among our research universities.

## NOTES

Thank you to Woody Powell and Jennifer Croissant for helpful comments on an earlier draft. I also wish to acknowledge the research support provided by NSF grant #9710729, The Association for Institutional Research Grant #99-129-0, and a small grant from the Social and Behavioral Sciences Research Institute at the University of Arizona.

1. Throughout this paper I use the term *privatize* specifically to indicate universities' development of proprietary rights to basic science findings through patenting.

2. Patents convey property rights for seventeen years from the date of issuance or twenty years from the date of filing.

3. Of course success in grant getting is tied to success at publication. In most science and engineering disciplines peer review for publication is single blind with reviewers aware of authors' identities and affiliations (Chubin and Hackett 1990). Thus, the Matthew Effect might be expected to hold across grant- getting and publishing, because in any specialty area there exist a limited number of reviewers and because both grants and publications contribute to a scientist's (or institution's) overall reputation.

4. While it is a commonly used measure, patent volume is somewhat flawed. Not all inventions are patented, and not all patents represent commercially valuable inventions. Only a small number of patents return significant royalty income (Association of University Technology Managers, 1997,1). Various attempts have been made to establish the value of patents using renewal data (Lanjouw, Pakes, and Putnam 1996), citation measures (Narin and Breitzman 1995), and interviews with executives in high technology firms (Mansfield 1995; Rosenberg and Nelson 1994). The general consensus of these studies is that patent volume represents an important measure of innovative output and value generally and of private science activities within universities (Jaffe 1989) despite flaws in the measure.

5. The Carnegie commission recognizes eighty-nine R1 universities as of 1994. R1 rankings are awarded based on a number of criteria including level of federal support. My data set includes information on only eighty-seven R1 institutions because the University of California system, whose nine campuses boast seven R1 institutions, patents through a central technology transfer office and does not report patent data at the campus level. To get campus-level patent counts, I coded individual patents based on the city of residence of the first inventor listed. For most campuses this posed no difficulty, but the geographic proximity of the Berkeley and San Francisco campuses made this coding strategy problematic; thus, I have combined the Berkeley and San Francisco campuses into a single observation.

6. Not all patents are assigned directly to universities. In some cases patents are assigned to governing boards as is the case with the assignment of University of California patents to the state system's Board of Regents. In other cases, patents are assigned to foundations associated with an institution as in the case of the University of Wisconsin's patents, which are assigned to the Wisconsin Alumni Research Foundation.

7. I do not present detailed descriptions of 1990s two-year lag models because those models are identical to the 1980s models described below with one exception. Model six, the patent characteristics model for the 1990s, includes the variable measuring the number of patents jointly assigned to industry whereas Model three, the 1980s patent characteristics model, does not. In all other particulars 1990s models can be created from the 1980s models by changing the year subscripts from 1987 to 1997 and from 1985 to 1995. By the same token, I do not present detailed descriptions of models run at a ten-year lag, explanatory variables for 1985 regressed on patent volume in

1995, because the details of those are exactly identical to the three regression equations presented for the 1980s. Again, the 1980s models with two-year lags can be transformed into the ten-year lag models simply by changing the year subscript of the dependent variable from 1987 to 1995.

8. For these models and for models of the relationship between public and private science in the 1990s, I chose a two-year lag (e.g., effects of eighty-five indicators on eighty-seven patents) to allow for the lag between filing and issuance of patents. Thus, these models should tell us the importance of university characteristics at approximately the time of patent filing for explaining the successful issuance of those patents about two years later. A later model examines the effects of these same variables over a ten-year time period to examine the longevity of effects for each variable.

9. When variables are log transformed, their regression coefficients no longer represent the effects of unit changes on a dependent variable. Instead, coefficients represent the effects of proportional changes. Without log transformations, changes in the monetary variables in my model are too small to show effects in patent volume. For instance, a unit change of $1000 in Johns Hopkins University's grant income of approximately $300 million dollars is too insignificant to be detected in the dependent variable. However, taking the log of federal R&D obligations for Johns Hopkins compresses the variable's distribution, making regression coefficients represent the effects of proportional changes in the variable. Thus the coefficient of the log of federal R&D obligations represents the effect of a 1 percent change in the variable on the number of patents assigned to a university.

10. Two dummy variables capture university location on the east coast or west coast. Universities in the Midwest represent the excluded category.

11. A dummy variable coded one if a university is private and zero if public.

12. A dummy variable coded one if independent variables are reported aggregated to multiple campuses and coded zero otherwise.

13. Much of this data is available for the 1990s from the Association of University Technology Manager's yearly licensing survey of universities. I will use these data in future work to expand arguments about the role organizational learning-based accumulative advantage plays in establishing a stable private science stratification order.

14. For a more detailed discussion of patents as intellectual capital in the biotechnology industry see Smith-Doerr et al. (1999).

3

# Entrepreneurship in Technology Transfer Offices: Making Work Visible

## W. PATRICK McCRAY
## JENNIFER L. CROISSANT

### INTRODUCTION

Nearly all major American research universities have administrative units whose task it is to promote, facilitate the transfer of, and license university-developed technologies to industry. A survey of university technology transfer (hereafter TT) shows that more than 5300 licenses were granted by universities between 1991 and 1995 (Massing 1996). The number of patents awarded to American universities increased from about 300 in 1980 to nearly 2000 in 1995. More than 1900 new companies are believed to have been created as a result of university licensing since the Bayh-Dole act was passed in 1980.

The Bayh-Dole Act, 1980 Patent and Trademarks Amendments (Public Law 96-517), was designed by Congress "to promote collaboration between commercial concerns and nonprofit organizations, including universities." The intent of the law was to foster the growth of technologically based small businesses and nonprofit entities, including universities, by permitting them to own the patents derived from federally sponsored research. Public Law 98-620, passed in 1984, increased the rights afforded universities by Bayh-Dole and permitted them to assign their property rights to others. This paved the way for universities to license technologies developed in their labs to outside companies for a profit. The Cooperative Research Act of 1986 (PL 98-462) permitted universities and industry to collaborate without fear of anti-trust litigation.

Such changes in federal policy and the interest of business in university-generated technology have led to a significant restructuring in the environment

55

and in the reward structure of research universities. Universities have become "the creators and retailers of intellectual property" (Chubin 1994, 126). The concept of a university or faculty member owning and profiting from the knowledge and technology generated through research was once thought to be a "graying of the ivory tower." Now science, technology, and property, once viewed as independent entities, have become contingent on one another through the concept of intellectual property (IP) and the commercialization of knowledge (Etzkowitz and Webster 1995, 490–491).

In the effort to articulate a "thick description" (Geertz 1973, 3–32) of TT activities, McCray spent four months conducting more than three dozen interviews with administrators and faculty engaged in technology transfer activities at a mid-sized Research 1 university. Those interviewed included staff at the Office of Technology Transfer (OTT) as well as outside administrators and faculty from both the life sciences and physical sciences. He observed staff meetings of the OTT and shadowed office staff through a typical workday. These observations were supplemented by examination and analysis of relevant documents, such as the minutes of meetings, research agreements, contracts, database entries, office communications, financial statements, and official reports and publications

We examine the diverse ways in which the office does its work and makes this work visible to others, evaluating the perceptions that staff at the OTT and university faculty have of each other and the expectations that faculty have of the OTT. The shift to more entrepreneurial policies for univerisities provides an opportunity for studying the invisible work that supports these policies and demonstrates how OTT representatives act as organizational entrepreneurs. The OTT actively defines and delineates what intellectual property is through its interactions with other members of the TT community and creates a history or locus of institutional learning upon which later decisions concerning IP can be made.

After giving a brief account of the OTT as an administrative unit at the university, we describe the network associated with university-based technology transfer. The OTT is one participant in this network which includes university faculty and administration, private business representatives, and elected officials. The OTT is at the center of a variety of different tensions imposed on it, because the other participants in the network have differing expectations of intellectual property and technology transfer. The interface between the OTT and university faculty is especially important.

## INTRODUCTION TO THE OTT:
## PERCEPTION, ACTORS, AND NETWORKS

The OTT was created at this mid-sized state university in 1988. In comparison with other university TT offices, OTT the OTT is relatively new. For example, the University of Wisconsin's OTT was created in 1925. With the passage of Bayh–Dole and similar legislation in the 1980s, a number of state universities, including Michigan, Florida, North Carolina, Washington, Pennsylvania, Texas, and Arizona, established TT offices. Prior to 1988, the institution had arrangements with outside companies to provide patenting and licensing services. In 1990, the university began to take a more active role in licensing agreements. The staff was expanded to three full-time employees including a director and two staff positions specializing in patents and contracts. In 1996, the staff expanded to include two licensing specialists in the physical and biological sciences. More recently, a database/office support person (1997) and a financial specialist (1998) were added.

The OTT crafted key policy documents in the late 1980s and early 1990s. These included the Board of Regents Patent Policy, a Conflict of Interest Policy, Invention Income Distribution Policy, and an Intellectual Property Policy. The OTT reports directly to the university's vice president for Research. A faculty advisory committee meets monthly to discuss and review office policy.

The OTT interacts with a wide range of institutions, agencies, corporations, and individuals as it carries out its work. These can be divided into four basic groups: institutional, private, professional, and state. Institutional actors are members of the university community, including but not limited to: the OTT; other university administrative units which encounter IP (Office of Research and Contracts Analysis [ORCA] and Sponsored Projects Services [SPS]); university faculty (both as individuals and as various faculty groups including the Technology Transfer Committee); the Office of the Vice President of Research; university legal services and attorneys; the higher echelons of university administration (deans, department chairs); and other state and private universities. Private, extramural participants in OTT activities include citizens and citizen groups; the "public at large" as referred to by government and academic representatives; private, for-profit businesses that provide research support for university research and/or acquire university-developed IP; private law practices; private, for-profit companies that manage IP and TT; and lobbying groups. Professional associations and groups formed by persons involved in university TT include the Association of University Technology Managers (AUTM) and the Licensing Executives Society (LES). Finally, the state has various modes of interacting with OTT activities, including the State Legislature, government-appointed bodies (the Board of Regents most notably), and the Attorney General's office.

The OTT is the primary entity at the university responsible for matters concerning IP. However, as the above list indicates, it is only one of several actors that comprise the overall network. Clearly, some of these actors are more relevant than others and play a more direct role in affecting the policy and process of university-based TT. It is somewhat misleading but necessary to treat these categories of actors as relatively homogeneous. For example, the category "private industry" has numerous subdivisions and distinctions based on industry, firm size, and other characteristics; further, each business or corporation has a culture that differs slightly from others in the same arena (Kunda 1992).

The OTT interacts with all of the other participants in this network; each group in the TT network has its own goals, perceptions, and expected means for accomplishing these goals. In comparing these goals and OTT's own goals and self-definitions, it is clear that there is considerable potential for conflict to arise. For example, how can the desire of private industry to obtain as much IP for the least cost be balanced with the need to protect state-funded IP? Or how does one balance the OTT's desire to make money for the university with the responsibility of the OTT not to take unreasonable risks (the very type of risks that businesses take every day)?

## The OTT and University Faculty: Perceptions of the "Other"

The themes of tension and conflict have appeared in other sociological studies of university-industry collaboration. For example, Rhoades and Slaughter (1991) examine the interaction between university faculty and administration in the negotiation of TT while Slaughter (1993) depicts some of the changes that have appeared in the narratives told by university presidents in the justification of the institutions' science policy. Campbell (1997) explores this issue further and identifies specific examples of potential conflicts: conflict of interest, conflict of commitment, and conflict over internal equity (359–361). These tensions represent a new type of conflict resulting from the increasing pressure to commercialize university-developed IP. In contrast with prior studies, the use of interview and participant-observation evidence permits the examination of tensions and conflicts as they occur between staff of the OTT and individual faculty and business representatives.

Faculty invention disclosures represent an essential, and often initial, form of IP. Faculty are one of the groups with which the OTT staff spends a great part of its time interacting. The OTT's mission statement indicates that providing service to the faculty is a central function of the office. We explore the perceptions these groups have of each other, focussing on three different domains: the OTT's view of the faculty; faculty members' perceptions of the OTT; and the OTT's view of itself.

## OTT Perceptions of Faculty

In his very first interview with an OTT staff member, McCray asked, "How do the faculty interact with the OTT?" This person laughed and replied, "Us?

Most of the faculty don't even know we exist. And this makes our job diffi-cult." Later interviews reinforced the perception that the staff believe faculty have a nebulous understanding of the OTT's purposes and goals, if they have one at all. For example, all university departments ostensibly have a copy of the "The Red Book," a manual assembled by the OTT which details faculty/staff responsibilities and rights concerning IP. However, it is not at all certain that this manual is ever referred to or even that university faculty know of its exis-tence. One faculty member in the sciences, when asked about this book noted, "I've never seen it and I'm not even sure we have a copy in our office."

The perception of university faculty with respect to their understanding of IP issues is an excellent illustration of the tensions that exist between the cul-tures of academia and the OTT. The basic impression of the faculty given by OTT staff when dealing with IP topics is that faculty are very naive when it comes to dealing with business and on IP matters in general. This perceived naivete can often lead to unrealistic expectations on the part of the faculty. Their unfamiliarity with IP issues can result in two specific types of conflicts.

First, university researchers, because of their unfamiliarity with the busi-ness world, may have a mistaken impression as to the commercial value of their research. As one OTT staff member stated, "Half the faculty don't even know what industry is." In some cases, PIs either over- or under-estimate the value of their work. Under-estimation benefits companies interested in supporting fur-ther research or acquiring IP rights. For over-estimation cases, it is OTT's task to explain to a disillusioned PI why their invention might indeed be very inter-esting but still lacking commercial viability.

A second conflict seen frequently in the daily interaction of the OTT with faculty stems from faculty misperceptions of how the technology transfer process works in the university setting. Faculty are inexperienced in the process and unaware of the amount of time it takes to negotiate contracts and licensing agreements. As a result, they can quickly become frustrated. Several OTT staff noted their own personal experiences of faculty contacting them on an IP issue at the very last moment and expecting the OTT to resolve their problem imme-diately, despite its being a complex legal issue. As a result, the faculty, according to one OTT staff member, begin to perceive the OTT as slow, adversarial, inef-ficient, and uncooperative. Researchers begin to view the entire technology transfer process as too complicated and perhaps not worth bothering with.

An important point noted by a staff member is that the faculty carry out their research under an "academic collaboration model." This implies some base level of mutual trust between partners that does not have to be detailed in legal documents. The fact that faculty apparently use such a model in their dealings with business was seen as another sign of faculty naivete. According to the OTT's director, researchers have to be made aware of why a "handshake between two parties" is not enough to arrange a research agreement. Faculty were described as failing to see why, despite the fact that thousands of dollars are at stake, lawyers have to be involved and why the process takes so long.

The faculty's concern with garnering research funds to be used in the immediate future typically exists in opposition to OTT's concern with issues of a much longer time frame. An OTT licensing associate (LA) explained that while he understands the PI's concern with getting research funding as soon as possible, there are broader concerns that need to be dealt with. The OTT, he explained, has a larger responsibility to the university that goes beyond the individual PI. Research and licensing agreements typically span several years, and the long-terms costs and benefits need to be considered. As an LA noted: "My job is to prevent loss." Faculty are described as excessively concerned with short-term benefits, such as obtaining research funding, while lacking a larger appreciation for issues that become manifest over a longer time span. For example, the OTT was negotiating IP and licensing for an invention with market potential in the hundred-million dollar range. The university would receive 1 to 2 percent and the inventor $50,000 to continue research funding. Faculty, according to the OTT, have a reputation for entering into research agreements that might give them immediate funding while sacrificing rights to IP in a manner that is unacceptable to the OTT. In this case, the OTT licensor noted: "[The PI] doesn't care about the agreement or whether I'm taking 2 or 3 percent, he wants his $50K."

Faculty in the College of Engineering were singled out by one LA as being especially guilty of entering into research agreements that sacrificed IP for the sake of "bringing money in today." This general tendency was explained by another OTT staff member as one of the "consequences of doing research in today's competitive environment." He went on to point out that, while faculty may be concerned with long-term IP issues in theory, "their focus isn't there," they are too caught up in the short-term aspects of their research, and "Most of them just aren't savvy enough to think it over for the long-term."

Part of the perceived naivete held by university faculty concerning IP was traced by one OTT staff member to geographic location. The university was described as not being located in a "strong business environment. Faculty just aren't exposed to these types of issues on a daily basis." Faculty in places such as California and Boston are thought to be much more sophisticated in IP matters. Part of this was ascribed to a longer history of university-industry interaction in places such as MIT and Stanford. Another perception was that the university is in a "cultural backwater" where cutting-edge businesses and the concerns associated with having them nearby are "just not in the forefront of the faculty's consciousness."

Understandably, the primary concerns of the university faculty, as articulated by the OTT staff, are thought to be obtaining funding for research and publishing the results of their work. This is especially true, as one of the OTT's legal advisors noted, in the case of untenured faculty. These PIs are seen as even less likely than other faculty to be concerned with long-term IP issues because of their focus on the immediate goal of obtaining job security. Predictably, there is also a gap between older and younger faculty in their views towards the utility and necessity of obtaining patents on their research. Younger faculty are perceived as having a more positive view of patents and the "commercialization of academia." One staff member noted, "People within the university treat patents differently based on their own view of patenting and how useful it is [in getting tenure]. There is a gap between newer and older faculty." Patents are also something that younger faculty can use as "bargaining chips" when negotiating with industry for research funding, possibly creating the situation where long-term IP rights are sacrificed for short-term research money. The tensions existing between faculty and the OTT were summarized by one staff member who noted that the main problem the OTT has with faculty is due to "a lack of communication and a failure of the faculty to listen. That's a bad combination."

## Faculty Perceptions of the OTT

Consider the OTT-faculty interface from the other perspective: how do university faculty view the OTT and university involvement in TT? Exploring the interface between university faculty and the OTT from the perspective of the faculty leads to interesting contrasts with the picture described above. One part of this contrast stems from the varying views toward IP held by university researchers; not all faculty view the increasing commercialization of the university in as sanguine a fashion as administrators of technology transfer do. One tenured engineering professor disapproved of the university owning patents and of overall trends in commercialization: "For the most part, I'm not in favor of what they [the OTT] do. Because our mission is education, research, and service. If we can somehow tie our research into industry and get sponsorship from a company, I don't know why we spend so much time and money in trying to tie up the rights." Moreover, he saw no reason why industry should not be able to take university-developed technology and do what they want with it: "If we're developing so much technology that companies are coming in and taking it, then we must be doing something right. That's one of our missions. If the company becomes more profitable, they'll hire more people, more people paying taxes. Everyone wins." Another engineering professor echoed this comment, stating, "For me, the OTT represents the interests of a drive to accomplish something the university should not be involved with: technology development. We should not set ourselves up as competitors of industry." When asked about the OTT's professed mission of getting technology out to the "public," the first engineering professor replied, "Leave that to industry. If we develop something really good, people from industry will pick it up." This faculty member was essentially articulating a model of academic collaboration that is very different from that adopted by other faculty as well as the OTT. This model was one in which the university would do work for a company which, in turn, feels obligated out of an ethical concern to put money back into the system. Legal arrangements would not be as important, and there could be other ways the university would benefit besides making money, such as hiring graduates, making donations to the university, and training students.

When asked if he really believed companies would invest money back into the system if not forced to through the licensing process, he said, "I think they would in many cases. The way these collaborative things are arranged is on a personal basis between faculty and someone at the company. I think all this legal stuff is not the way to go." This approach seemed to ally the faculty more with interests and concerns of private business. As he explained:

> My understanding is if there is something that is patentable, the university wants to own the patent. If they offer it to the sponsor, they will have first rights to it, but the university still will own the rights. This gives a lot of companies heartburn. They don't understand why they're paying for something they don't get. My attitude is that if the company can profit from it, then so be it. Maybe I'm naive but I think that is the way we serve society.

Faculty do not share these perspectives. A tenured professor in the life sciences heard the above model of university-industry collaboration. His response was: "That's naive. Technology transfer is definitely needed. The legal part of it is to protect scientists. They [the OTT] know the expertise, the terminology to write licenses." These models of university/industry collaboration were rather divergent. The picture of academic research in the life sciences was described as a very competitive one in which "You have to protect yourself from fellow researchers. The life sciences are very competitive. A patent protects your product while you're working on it." It is especially telling that this particular professor referred to his research as "the product." Moreover, this person felt the OTT should be even more aggressive when it came to the university profiting from faculty research: "They would reap the benefits, maybe bring in more personnel to help. I'm really surprised they're not thinking more in that direction. If they're not, they're losing out on what they could be getting. It's a Research 1 university."

The general view held by the OTT that university faculty are naive when it comes to issues of TT and IP was seen, based on the interviews with faculty, to be more complex than this. Clearly, whether a faculty person was naive depended a great deal on his or her personal experience, research activities, and level of involvement in technology transfer. One faculty member in the physical sciences bristled at the suggestion, saying, "I would be surprised if that statement [faculty are naive] is true for the entire faculty. I know that it absolutely isn't true in this department. The problems are more likely to arise on the part of faculty who are trying to understand the position of the OTT." Another interviewee replied to the topic of faculty naivete by stating, "I think that's overstating the case. Some faculty are very astute at figuring out what industry does. I think there are also a number of incredibly naive faculty, so the answer is somewhere in the middle. Faculty, however, simply don't believe that their work is not their own. Contracts are a way to get money. They don't think about the trade-offs. It's gotten very money-grubbing." This person went on to elaborate: "Faculty are more interested in just doing the work. In a lot of cases, it's "where do I sign and where's my check?" That's the negotiation. I don't think they read contracts. If they read it and think about it, they think "how are they going to enforce this?" They never think about things like background rights. You could be signing away your life's work." This comment echoes points made previously by OTT staff that faculty are "just interested in getting their research money."

This point was disputed, however, by a researcher in the physical sciences. He disagreed with the picture of university faculty only being concerned with the short-term procurement of research money. While admitting that such situations probably did exist, he felt they were "fairly rare." He referred to the research work done by his own department and noted that it was usually connected to the needs of industry over a longer period of time. Citing this interaction with industry as necessary, he referred to what he called "interaction synergism" centered largely around the manufacturing needs of industry, and "without that industrial interaction there aren't going to be any patentable ideas in the first place." He described a picture in which the university researcher established a long-term relation with a company, carrying out applied research that was largely in line with that company's research interests.

This is a much different model of collaboration than one described by a tenured professor in the life sciences. He categorized his research as "basic science" and felt that his work and needs with respect to the OTT were very different from other researchers or departments that are more problem oriented.

> I think all universities try to hang on to it [IP] and put things in writing with the hope that they will get something. In most cases, it never comes to light. I think most of my colleagues are like me, trying to do some science and get their lab funded and not really into [personal] money other than research money. Departments like [ours] can't be run the same way as civil engineering or optical sciences. They're developing new technology. We're testing concepts, not for the objective of making better plants but just really basic science.

There clearly was a variety of views held on the purpose and practicality of IP and university TT. Some of this difference resulted from whether one was in the life sciences, physical sciences, or engineering.

That the OTT was frequently seen as an impediment by the university faculty was acknowledged by the OTT staff. For example, one staff member said that he was aware that faculty frequently saw their relationship with the OTT as "adversarial," explaining that he felt this was more prevalent in certain research areas such as engineering. Because engineering faculty have historically had a long history of close collaboration with industry sponsors, it was generally felt that engineering researchers resented the intrusion of the OTT into relationships that were already established. This staff member felt one of the difficulties the OTT had to surmount was the perception by faculty that the office is an "obstacle."

## The OTT's View of Itself

A final perception to consider is a reflexive one: how does the OTT perceive itself as it participates in TT activities? What does the OTT consider its role in the TT process to be? Based on data collected in interviews with OTT staff, making a profit was not seen as the primary function of the office. Instead, other motivations, such as serving the faculty, protecting the university, and developing technology for the public good, were heard often. Staff also noted that the manner in which business was conducted at the OTT would be very different if profit were the sole motivating factor.

While all OTT staff described the primary function of the office as "providing service," there were conflicting thoughts about the OTT as a profit-making enterprise. One of the licensing executives described the OTT as primarily a service organization and not profit-driven. Yet he noted:

> We're definitely a business culture because our whole objective is legal, business, creating agreements, contractual obligations, etc. All of this is really more of a business orientation. But what we're working with is academic knowledge, some of which is patentable and can make money. The general theme was, in the old days, universities would create knowledge and companies would take the knowledge and go make money on it. Now the theme is create knowledge, some of which is commercialized. Why can't universities make money on it?

The varieties of TT offices were frequently described in two basic categories. One type was "profit driven" with Stanford and MIT given as examples. The second type was "service oriented" viewed as the role of this university's OTT. The distinction seems to correlate with whether the university is a private or public institution. The goals of the OTT are "to keep the university out of trouble and keep faculty happy." However, this mission was further expanded to protect the university and the VP of Research from the legislature and "not to sign agreements which contravene state law and not to give away stuff." Money was "not an overriding factor" in this scenario. While the OTT was again seen as providing a service, staff did concede that the office "has to walk a line between the public good and the desire of the involved companies to make money." Because of the partial-profit motivation, the office is then placed in a situation where it handles its affairs differently from a TT enterprise that is solely motivated by profit.

For instance, staff members noted that, because of their service orientation, they often handle disclosures and IP with no obvious potential. In a purely profit-driven enterprise, such disclosures would be returned to the inventor with no action taken. Ignoring obviously noncommercial disclosures would reduce the workload at the OTT and increase the efficiency and turnaround time of the office. However, this cannot be done without violating its service directive. Therefore, the OTT handles disclosures in a manner different from a private company, with the result that additional conflicts emerge.

The OTT staff perceive themselves as performing a range of functions that are different and yet interrelated to other university offices that also manage IP issues. The OTT is responsible for any IP issues that arise in the other offices. All of the offices involved with research administration at the university look at agreements differently. However, each has a different function; ORCA was identified as the "gatekeeper," and SPS handles post-contract awards ("a bunch of accountants, so they have a different orientation," according to a OTT staffer); the OTT's function is to handle IP.

## FACULTY EXPECTATIONS OF THE OTT

Given that the OTT's primary function is to provide a service to the various members in the TT network, each of which has expectations of the office, we again turn to the expectations university faculty have of the OTT.

### The OTT as Facilitator

Timeliness and efficiency on the part of the OTT are essential components to one of the key expectations that faculty have of the OTT; faculty see the OTT as offering a service and wish to procure this service easily. In this respect, the OTT often acts as a facilitator for faculty research activities. For example, a Biological Material Transfer Agreement (BMTA) was hand-delivered to the University's Cancer Center, instead of through campus mail. The most common faculty complaint was that the OTT was too slow in responding to the needs of university researchers. The system is apparently working much more smoothly now than three years ago, before the OTT hired new licensing associates. Still, a researcher in the life sciences noted "that there still is quite a lag phase involved. I guess it depends on what the university wants to put into it [TT] in terms of what they get out of it. I think if they manned the office the way it should be manned, they would be reaping the benefits from it." Another researcher noted the linkage that exists between publishing and patents, which requires the execution of the OTT's work in a timely manner: "I am paid by publication not by patents. Patents are my hobby. Nobody really pays attention to what I'm doing unless there are publications. In some ways, this is a conflict of interest because you have to publish on one hand, and on the other you are not supposed to publish until the patent is finished. This is what always makes me nervous. A timely fashion [for the OTT to get patents processed] is very important." Over the broad spectrum of research activities carried out by university faculty, there was considerable variability expressed as to what it meant for timely action on part of the OTT. A large part of this variability can be traced to the difference in work between researchers in the life sciences and engineering or physical sciences.

During interviews, OTT staff and faculty explained this phenomenon by referring to the corporate cultures with which different academic departments work. The electronics industry, for example, "may not have time to mess around with patents," referring to the considerable period of time it takes to be granted a patent in relation to the lifespan of a product. Businesses in this area may instead get a monopoly by market share rather than through a patent. Electronics companies were depicted as frequently cross-licensing patents with each other; patents are "bargaining chips" rather than means to establish commercial hegemony. An LA described the idea of patents in engineering and the physical sciences as "encumbering" and not something with which most faculty were generally concerned.

In contrast to electronics, a patent in the biotechnology industry may have greater value because of the longer lifespan of a product or process. More time is required to bring a product to market because of the need for regulatory approval. Faculty from the life sciences, especially those doing applied research, appear to be much more interested in obtaining patents. One researcher explained that this offered a way to protect one's research from others in the same field, including those who might be employed in the private sector: "Sometimes it might take years to develop that product, and during that time the disease you're working on might no longer be prevalent. A patent at least protects you while you work on the product." Or consider an agreement made in which the university received 60 percent of royalties from a peptide technology, while a technology management company received 40 percent. Now, university cancer researchers wanted to use the peptide technology in conjunction with an invention of its own, and the OTT was negotiating with a company for technology that was originally developed at the university.

Time is also perceived differently, depending on whether one is working at the OTT or conducting research. One of the LAs gave several examples of how faculty do not understand the intricacies of the process, the amount of paperwork involved, and the complications that arise in the legal negotiations. For example, in an office file for a licensing case in the life sciences, there were over one dozen copies of the same basic contract that had been sent between the legal representatives of the universities and companies involved. The process took over a year before a final contract was agreed upon.

Misunderstandings occur because faculty are unaware of all of the ancillary factors involved. One LA said, "The OTT may be a service organization, but faculty don't understand the service provided." This problem may even be more pronounced for younger faculty who are under pressure to get their research funded and published. For such persons, delays of several months may appear as career threatening instead of merely inconvenient. As one person noted, "My perception is that some people report to the director [of the OTT], who has to report to the VP of Research. This creates a timing structure in which it is difficult to get things done in a short period of time." Another aspect of faculty's expectations that the OTT will facilitate their work is "efficiency"; faculty expect the OTT to manage IP and TT issues in an orderly, competent, and streamlined manner. Another researcher, whose primary interaction with the OTT also involves BMTA's requested by private companies, complained that the processes used by the companies and the OTT were too slow and inefficient: "One problem is that offices of tech transfer, say at Oklahoma State and here, have slightly different wording that gets them walking in circles for another four weeks until they have it sorted out. I guess it's just the way of any administration of generating enough work so their positions can be justified. . . . For me, it's just a waste of time." The researcher questioned if agreements, such as BMTA's, which he saw as fairly routine, might be exempt from signatures and not routed through the OTT: "Give me a sheet of paper I can attach." He also noted that the amount of time he spends on handling BMTA's (which primarily involve companies requesting gene sequences he felt they could generate themselves) is taking up an increasing amount of time. As a result, he bypasses the OTT in some cases and sends the materials out himself to save time. This is just one example of the multiple paths taken by university faculty to facilitate, control, or avoid interaction with the OTT.

In another case, the slowness of the process was cited by an interviewee as actually preventing the successful completion of a planned research project. The faculty person felt that the office didn't respond fast enough and that "there were several iterations of paperwork between the OTT and [the company]." All delays were, according to the researcher, centered around IP issues; finally, it was too late, the company pulled out, and the researcher and the university lost the contract.

Staff at the OTT and university administrators also acknowledged the problems and perceptions that faculty have with the TT process. A member of the university's Technology Transfer Committee noted that the most common faculty complaint was that the OTT was too slow and inefficient. According to him, an external review of the office made a few years ago noted, "The OTT needed a minimum of four new people to function efficiently. They hired two. The problem boils down to the fact that they simply don't have enough people to get the job done. One can also argue that the whole structure is wrong, and simply adding more people won't fix the problem." One part of the problem of perceived slowness and inefficiency was cited as existing higher up in the administration, specifically in the Office of the Vice President of Research and his own particular views of the TT process.

> His strategy is to please people, to keep as many people happy as possible. He keeps the major squeaky wheels happy, but in terms of the rank-and-file faculty he's not. He has to decide who's in charge and how it [TT] is actually going to happen. He needs to define pathways instead of letting them be self-defining, in the sense of "if I want to do something, I have several options available to me." The root of the problem is that the university grew too quickly. You see that reflected to some extent in the Byzantine structure which is one way this university has attempted to solve the problem. A good bit of the problem is that the faculty will follow the path of least resistance.

The expectation held by the faculty that the OTT should be efficient also extended to include the profitability of the office and what these profits should be used for. These thoughts were expressed, for example, by a faculty member in engineering who felt that the OTT should definitely be able to support itself and that making money for itself should be

> a basic criteria for their continued existence. You must know that after learning the office ran in the red for years, the fact that it expanded was a mystery to many people. It's a difficult nut to crack. Because they [OTT], based on their mission, stand in the way of faculty getting contracts, it's hard to see them as providing a service. The notion that they're there to make a profit is at the root of their difficulties. The perception is that there is this home run that they are trying to pursue, and as a result they place us at odds with potential sponsors. That's where I see the biggest problem. And again, the roots might not be in the OTT. There might be a lack of proactive involvement by the upper administration to settle IP issues, to create official ways that a company who wishes to obtain IP can do so without it being prohibitive. It all stems from the VP of Research, I guess, because he got burned once and now is very wary in contract negotiations.

However, one of the LAs disagreed with the idea that university faculty want the OTT to be self-supporting:

> I'm not sure I believe that statement. I'm not sure they [the faculty] believe that. They disclose here and regularly they ask, "Are you going to patent my invention?" and I say we're looking for a licensee to cover the patent cost. They get very disturbed and think we should patent their invention. If you explain that we have a 150 inventions a year, and it's a minimum of $5,000 to patent in the U.S., that's a lot of money. Where's that money going to come from? I don't think anyone says the university ought to fund that. Yes, we agree that we should patent in an ideal world, but I think the faculty don't really agree, they want to bring money in, and they want the invention patented. I mean, faculty are great ones for spending money. If this were a corporation, we would focus on that, and things that aren't bringing in money we would just dismiss. The other thing is that income takes time to generate. Most TT offices take time to be self-supporting. UC became self-supporting in 1989. I was there at the time. The office had been in red for about twenty years before it became self-supporting. So, yes, everyone wants this office to be self-supporting, but it's going to take time to get there. California poured lots of money in. I think this university is in a stage where it's costing money, but we will be self-supporting at some future point. The question is how to get there.

These faculty comments point to some of the tensions and conflicts inherent in the process of university-based technology transfer. These conflicts strongly suggest that the goals and criteria for successful TT vary, depending on whether one is a university researcher or administrating the products of this research.

## The OTT as Protector

The OTT acts in a "protector" role. Consider the example of a meeting arranged for a faculty member, Phil, whose primary concerns centered around obtaining protection from the OTT for an invention. The company sponsoring his research wanted to lay claim to the IP, although the money was provided in the form of a gift, not a contract. What did Phil, the PI, expect to obtain from the OTT at the conclusion of this meeting? First and foremost, Phil was interested in protecting any rights that he had with respect to intellectual property developed by his students and him. While he expressed concern about losing a valuable source of research funding, he (and certainly his students) were more interested in their ability to continue the research after obtaining funding from another source and being able to publish their results. The university attorney mentioned to Phil that, if a license was granted on the technology, there was a possibility of the investigator and students getting some royalties. Based on their expression of surprise and pleasure, this was not something they had considered. The OTT director was concerned that the university "can't give away technology like that." Finally, Phil was interested in making sure that his soured relationships with the original funding source did not create problems for his future research—in his words, he wanted to make sure that "the company did not screw him or his students." In short, Phil's main expectations with respect to getting protection from the OTT for his IP were to: (1) maintain the ability to continue doing research and to publish, (2) safeguard the IP that had been developed, and (3) prevent the funding company from making any unreasonable claims to it. While not a central interest, there was also the possibility that he and his students might be able to make a personal profit on their IP.

## The OTT as Educator

In addition to facilitating faculty's participation in the TT process and providing protection with respect to IP issues, faculty also expect the OTT to provide education and instruction on university IP policies and protocols. Consider the following scenario.

Alex, a faculty member in engineering, requested a meeting because he had some questions concerning certain aspects of the university's IP policy. Alex was, at that moment, writing an NSF grant which pertained to the reform of engineering education. His main question revolved around copyright issues and the ownership of IP (Web sites in this case) developed by university faculty. Was creating Web sites the same as a faculty member writing a textbook? As publishing technology changes, new situations are occurring that are not explicitly covered by university OTT policy. Celia, the OTT director, explained that the standard OTT policy was that the university could lay claim to royalties but doesn't because of a tradition of letting faculty keep that revenue stream. She further noted that this could all change, and the university, in addition to revising its overall IP policy, was going to have to formulate new terms for what she called "novel publishing methods."

Much of the confusion and source of disagreement between Celia and Alex centered around the issue of what is "traditional academic work." Policies for the new developments in electronic publishing are not clearly defined. Alex wanted the OTT to clarify this grey area and mark its boundaries, as well as to define the borders of what "traditional" publishing was.

This meeting was arranged at the faculty member's request. The primary topic revolved around Alex's interpretation and understanding of university IP policy. Due to vagueness in the policies and the rapidly changing circumstances of Internet publishing, Alex wanted clarification. The OTT, represented primarily by the director, attempted to clarify existing IP policy and educate the faculty member. At the same time, as OTT developed and refined policy with respect to a particular form of IP, copyright ownership, they articulated a second function in not only clarifying definitions and educating faculty, but in actively constructing the definitions. Changes in practice bring new technologies and ideas into view as potential IP, and the OTT enlarges its boundaries and acts as an entrepreneurial agent in defining IP.

## WORK AT THE OTT

### Defining IP

In a meeting with the OTT's IP management specialist and the LA for the physical sciences, the new head of university research strategy at Hi-Tech Inc. (Norm), explained that he was making a tour of different universities to meet with TT staff to develop a "strategic approach to TT" by "relating university needs to Hi-Tech Inc.'s needs." He passed out a handout to everyone, then used this for the next twenty minutes to lay out his case. Norm explained that Hi-Tech had identified two potential PIs who were willing to do work with the firm. He wanted to put in place a basic research agreement that had good IP terms. This, he said, would facilitate the process of creating more agreements with the university in the future. Norm noted that Hi-Tech had about $1 million (a number which didn't seem to faze anyone at the OTT) to spread around for research, and that they were interested in setting up grants to University PIs of from $50 to $100 thousand over three years.

Norm's major concern was to get agreements in place as fast as possible. His desire was to distribute the money to faculty within 120 days once the company and the university worked out a standard agreement. Hi-Tech had created a standard research agreement with another state school, and it wanted to use this as a template to develop a standard agreement with other universities. He now passed around copies of the agreement to everyone. Celia was in favor of Norm's basic idea, as it would help clarify and smooth out the TT process. To her, the Hi-Tech agreement looked good, but it needed to be gone over carefully. Celia pointed out to Norm that the mission of the OTT was twofold: to train students and to get technology "out there," and this agreement would facilitate that.

Norm's agreement spelled out what was meant by "IP" and how the division of ownership was to be constructed. It defined and determined the limits of a standard IP arrangement between Hi-Tech and the university, and it put a box around what the IP is and what the financial benefits and costs might be to the parties involved. In this case, the concern with doing research and making agreements in a timely manner for the microelectronics industry was evident.

Staff define IT through negotiations with faculty and firms. This includes activities such as creating the terms of a license or research agreement, evaluating a disclosure's commercial potential, deciding to file for patent, and refining OTT policy with respect to certain types of IP. This is where one of the powers of the OTT lies: in defining and negotiating the content and terms of IP.

One root cause of the tensions between the OTT and the faculty is the variance between differing views that faculty, administrators, and legislatures have of intellectual property. IP, in this case, is a "boundary object" (Star 1989; Star and Griesemer 1989). A boundary object

> inhabits several intersecting social worlds and satisfies the informational require-
> ments of each of them. Boundary objects are . . . both plastic enough to adapt
> to local needs and the constraints of the several parties employing them, yet
> robust enough to maintain a common identity across sites . . . they have differ-
> ent meanings in different social worlds but their structure is common enough
> to more than one world to make them recognizable, a means of translation.
> The creation and management of boundary objects is a key process in devel-
> oping and maintaining coherence across intersecting social worlds (Star and
> Griesemer 1989, 393).

Boundary objects are created when different social groups have varying visions and interpretations of the same thing. IP has a different set of associated meanings depending on whether one is creating it or managing it. IP is the representation of an invention, be it a device or a process. It may even be thought of as a disembodiment of the invention. In addition to referring directly to the invention as a specific object or process, it also assumes a somewhat more abstract character as "intellectual property".

These scenarios between the OTT and the representative from Hi-Tech, the faculty member who was interested in copyright issues for Web sites or the corporate claims to work, represent situations where the OTT is defining the terms of IP and making decisions about what is to be included in the "IP box." In their interactions with university faculty and industry representatives, the staff at the OTT mediates between these two groups as to what constitutes IP. In doing this, the OTT helps create a consensus on what, in reality (as represented by the terms in legal papers), the content of the IP is. The general focus of these discussions is the same in each case—defining and delineating IP. This activity emerges as one of the OTTs most important functions.

Two major kinds of tools are employed by the OTT in the process of defining the content and boundaries of university-developed IP. These include formal legal papers, various modes of mediation, and the knowledge and judgment tools of IP specialists. One of the products of the OTT is an array of legal documents concerning the terms, limits, and definitions of IP. These include faculty disclosures, provisional patent applications, patents, research agreements, licenses, and contracts. Each type of legal documents is a congealed representation of negotiations and consensus building carried out by the OTT to explicitly determine and identify the nature of intellectual property. These legal papers are the outcome of a series of communications and mediations between staff at the OTT and other participants in the technology transfer network such as businesses and faculty members.

An equally important tool in the mediation of IP is the need for the LA to use his or her personal experience and tacit knowledge to define and delineate university IP. One situation in which this judgement is most necessary is when LA or the office director must evaluate a faculty invention disclosure for its commercial potential. University faculty submit disclosures to the OTT whenever they have an "invention," however loosely that might be defined. The funding source of the research that resulted in the disclosure is verified. At twice-monthly meetings the OTT convenes to manage new disclosures and assigneds them to a "manager" from among the office staff. Generally, disclosures are divided into two categories: Life Sciences and Engineering/Physical Sciences. At the next disclosures meeting, the status of the disclosure is summarized by the manager, who makes a recommendation as to what should be done with it.

The manager of the disclosure evaluates the "invention" described in the disclosure for its commercial potential. Both OTT's associates have industry experience in their respective domains (Life Sciences or Engineering/Physical Sciences). The process is simplified in some cases, as the university PI may already have a specific interested industry sponsor in mind with whom he or she has a relationship. Determining a disclosure's commercial viability is largely the personal responsibility of the respective LA. It is based primarily on subjective feelings about the invention's commercial potential and past experience with the PI, the type of technology, and the possible companies involved. It is easy to imagine how conflicts can arise between the OTT and a faculty member who has filed a disclosure only to find out several weeks later that it is not thought to have any marketability. Under these circumstances, the invention is usually "released" back to the PI. In some cases, disclosures are sent to outside firms when it is very difficult to evaluate their potential or to demonstrate that both the university and an outside firm have evaluated the disclosure and found it lacking market viability.

The timing and content of a disclosure constitute one way in which faculty attempt to exercise control over the TT process and, in effect, draw a line around IP and declare what they perceive it to be. The director of the OTT described this as a potentially serious problem and a source for conflict between the office and the faculty. In fact, the early filing of a disclosure which was subsequently returned to a university faculty and developed further was a contributing factor in a substantial lawsuit filed against the university several years ago. The director noted that the practice of filing a vague and premature disclosure form occurs occasionally and ". . . is one strategy to keep the OTT out of the process . . . I don't like to return disclosures . . . we get a set of disclosures, for example, that have a sentence each and they're asking us to return them as soon as possible." A sociologist at the university, who has also examined the practice of university-based TT, concurred, saying that this is one way in which the faculty is attempting to gain control over the process by choosing when and what to disclose (Rhoades, personal communication, 1998). This view was also confirmed by a member of the Technology Transfer Committee, who noted:

> A faculty member files a disclosure, and the faculty member wants it returned to him . . . what the faculty member does is say, "yes, it's a disclosure," so the OTT looks at it and there may be, there has been in the past, some collusion between deans and department heads, so they'll [the OTT] say "this thing is not financially feasible" and return it to the faculty member, who then goes and does additional work without telling the university and then licenses it himself.

Clearly, the ability to "draw a box" around IP—to declare what it is and to define it—is very powerful. The fact that different groups in the TT network have differing perceptions and interpretations of IP (i.e., where is the box drawn) can be a source of tension and conflict. The activities of the OTT are directed, to a large degree, towards defining and delineating the contents of the IP box using a variety of tools: meetings, phone calls, e-mails, letters, legal documents, and the tacit knowledge of the staff. But defining IP alone is not enough to justify the involvement of the OTT in university TT activities. The OTT must also find ways to showcase its role in technology transfer to the other participants in the TT network.

## CONCLUSIONS: MAKING WORK VISIBLE, BOUNDARY SPANNING WORK, AND INSTITUTIONAL LEARNING

One comment heard frequently during interviews and observation sessions at the OTT was that the university faculty does not understand the role the office plays in the transfer of technology. OTT staff reported that they thought the faculty were inexperienced in the process and unaware of the amount of time it took for the OTT to negotiate contracts and licensing agreements. As a result, the faculty quickly became frustrated. Moreover, communications between the faculty, as a whole, and the OTT were depicted as not optimal. A good example is the comment by one staff member who noted that the main problem the OTT had with faculty was "a lack of communication and a failure of the faculty to listen. That's a bad combination." Almost all of the OTT staff interviewed indicated that improved communications between the office and faculty would benefit both sides and that, ideally, there would be a greater amount of time spent educating the faculty about the OTT's procedures and activities.

While it is not explicitly noted in its mission statement, informing the faculty and the university community at large about its role in TT is one of the OTT tasks. This activity falls under the larger category of making the OTT's work visible. For example, during one of the shadow sessions, an LA printed out a list of his activities for FY98 that was compiled for the director and indicated the types of activities in which he had participated. It included ninety BMTA's, twenty-three Confidential Disclosure Agreements, and thirteen Licenses and Options, with a new agreement being negotiated about every two days. This type of information is included in the OTT's annual report sent to the VP of Research and to others. These reports include such information as money generated from agreements, number of materials transfer agreements negotiated, patents granted, patents in negotiation, and so forth. In short, they provide an overview of the TT activities that staff participated in that year.

Visibility is established for the benefit of the OTT staff and for its constituencies throughout the university network. The invisibility of OTT work is perhaps one reason that university faculty do not understand the TT process. Unless the faculty are intimately involved in TT activities, the work of the OTT appears to them as a "black box." Why is making the work of the OTT visible an important activity? On the one hand, the reports generated by the OTT that summarize the year's activities and successes are a way of rationalizing the existence of the OTT. Sending such reports to the VP of Research and making them available for others to see affords the OTT a means to present its work, albeit in a very abstract form. These tools for showing the work of the OTT play a role in justifying the existence and continued support of the OTT by the university administration. Such reports, which do show an increasing level of activity in almost all areas from FY93 to FY98, could be used by the OTT to ask for continued or increased availability of resources.

On the other hand, the process of making work visible is important in the daily running of the OTT. It affords a means of creating a record of the office's activities. By creating records such as disclosure and agreements databases, the OTT is, in effect, creating a historical record of its work.

The process of defining IP, which is a multivalent concept or boundary object with differing meanings to the actors in the TT network, is a powerful tool of the OTT. The process of defining IP through mediation and the creation of consensus is also a fundamental cause of tension between the OTT and members of the faculty and business communities. While the act of making their work visible is important in communicating to other actors in the TT network, it is also a way for the office to internally create its own history. This history, in turn, becomes a resource for the continual refinement of OTT policy used in the daily definition and delineation of intellectual property. As Jason Owen-Smith (chapter 2) suggests, the ability of an OTT to be a locus of organizational learning is essential to the potential profitability of technology transfer activities for a university.

In many senses, the multiple roles played by the OTT inevitably foster some of the tensions between it and the other actors in the TT network. If it were run solely as a for-profit business enterprise, then many of its motivations would be clearer. But in trying to satisfactorily fulfill a range of different functions (provide service, generate profit, provide protection for the university), compromises are made and conflicts arise. Barring a major restructuring (probably through legal change) of the university TT network, it appears unlikely that the conflicts and tensions will be eliminated. These tensions both represent and constitute the boundary between the university and its external constituencies.

While scholarship has been extensive on the institutional dimensions of policy changes surrounding university commercialization activities, including historical surveys (Geiger 1988; Bowie 1994; Weinger 1986) and studies of norms, values, behavior, and conflicts (Etzkowitz 1989; Slaughter and Rhoades 1990; Campbell 1997; Seashore-Louis, et al., 1989), little work has been done which illustrates actors and mechanisms by which these changes are implemented and how different participants react. And while Guston (1999) discusses how the OTT operates as a boundary organization within the National Institutes of Health, he does little to flesh out the multiple perceptions of researchers and TT officers and the work that OTT staff actually do. OTT representatives both span and reinforce boundaries between the university and private industry. The OTT staff do not merely transport values and definitions among the different groups, but mediate and in some instances create working definitions that over time become increasingly stable. As agents who must speak to actors both inside and outside the institution's boundaries, OTT staff are remarkably powerful in defining IP through their active interpretation and implementation of policies and simultaneously remarkably dependent on the cooperation of the faculty researchers. They have been, until now, understudied actors in the system of university-industry relations.

## NOTE

The authors wish to thank those persons, both faculty and staff, who participated in this work by agreeing to be interviewed and observed and for patiently answering the most basic questions concerning technology transfer. Meredith Aronson, Teresa Campbell, W. David Kingery, Gary Rhoades, and Sheila Slaughter advised in the development of this work, while Dorothyanne Peltz was instrumental in the final preparations for publication.

# 4

# Harnessing a Public Conglomerate: Professional Technology Transfer Managers and the Entrepreneurial University

## WILLIAM N. KAGHAN

The modern American research university, in which research as well as teaching and community service are routinely expected, first emerged as a distinct type of organization during the last quarter of the nineteenth century. As the United States rose to the status of world power in the first half of the twentieth century, the research university matured (Geiger 1986). In the quarter century after World War II, when the U.S. was a global superpower engaged in the Cold War, the research university—with vastly increased government support—experienced a "golden age" (Geiger 1993). However, in the last quarter century, public/government support for research universities has become more qualified, and a number of "market-oriented" institutional changes have been initiated to try to improve the performance of universities in various arenas (Slaughter and Leslie 1997).

One particularly interesting arena in the most recent era has involved efforts to redraw the boundaries between university (i.e., basic) and industrial (i.e., applied) research in order to better support national goals such as economic development (Baldwin 1988; Guston and Kenniston 1994; Nelson 1995; Lee 1996). Prior to the Second World War, questions about appropriate collaborative research relations between universities and industrial firms were usually centered on regulating the movement of top scientists and engineers between university positions and industrial positions (Etzkowitz 1983). The transfer of university-generated technologies into commercial settings was quite limited (Matkin 1990). However, with the rise of "big science" (i.e., large-scale publicly funded research) after the end of the Second World War, the scope of what was involved in moving "knowledge" out of universities and into industry was much expanded (Mowery and Rosenberg 1989; Galison and Hevly 1992). Many of the most interesting institutional changes in university-industry research relations since the mid-1970s have involved efforts to rethink the "governance structures" (Williamson 1985; North 1990) set up to manage large scale interorganizational interactions between universities and industry. One particularly interesting institutional change has involved the emergence of specialized university technology transfer offices and professional university technology transfer managers.

    In this chapter, I examine the work that professional university technology transfer managers, who emerged in the late 1970s and early 1980s, routinely perform to mediate the transfer of university-generated inventions into more commercially oriented industrial research and development processes. I argue that this work revolves around the construction and maintenance of formal and informal understandings that underpin university-industry research relationships in such a way that university-generated inventions and know-how are more effectively transferred to industry while protecting the university's concentration on basic research. Although the role of entrepreneurial faculty in the process of technology transfer has been noted in the past (Brett, Gibson, and Smilor 1991; Etzkowitz 1997), I argue that the entrepreneurial university is more than an aggregation of unfettered entrepreneurial faculty members. Specialized technology transfer offices and the professional technology transfer managers who staff these offices play an important role in "harnessing" entrepreneurial faculty and the research efforts they are involved with, thus providing some coherent direction to the entrepreneurial university.

    In the first section of the chapter, I review conventional models of the relationship between basic and applied research and develop a new process model that better accounts for the work that specialized technology transfer offices and professional technology transfer managers do. In particular, I argue that conventional models are oversimplified and misleading because they focus too much attention on the research work conducted in laboratories and too little on "articulation work" conducted in offices (such as technology transfer offices) by "nonresearchers" (such as technology transfer managers). In the next section, I examine the work of specialized technology transfer offices and professional technology transfer managers in more detail. I discuss how the articulation work done in technology transfer offices serves to disentangle basic research results from unnecessary ties to the universities (viewed as "public conglomerates") in which they were originally generated and progressively entangle these results in research and development projects conducted in private industry. I label this work "arranging deals" and provide a model of the deal as an ongoing process. In the final section, I illustrate the issues discussed in the previous sections by telling three stories about the transfer of university-generated software applications based on the results of a three-year field study conducted at the technology transfer office of a major U.S. research university.

## OUT OF ACADEMIA: ORGANIZING THE EXPLOITATION OF DISCOVERIES ON THE "ENDLESS FRONTIER"

In the wake of the scientific and technological successes of World War II and the perception of basic research as a critical component of continued national development, the U.S. government took a more active role in supporting university and public sector research (Bush 1945; Steelman 1947; Kleinman 1995). During the 1950s, a number of government agencies, such as the National Science Foundation, the National Institutes of Health, and the Office of Naval Research were established to oversee funding of basic research conducted in universities (and other public and nonprofit research centers). In providing this funding and setting up these agencies, the federal government facilitated the move from "little science" to "big science" (Ravetz 1971; Galison and Hevly 1992) and more firmly incorporated academic research as an important element of the U.S. "national innovation system" (Mowery and Rosenberg 1993).

Along with these concrete changes in national science and technology policy and institutions, a new linear model of the relationship between basic and applied research was developed (see figure 4.1). The image of basic research that is incorporated in this model is best exemplified in the essays of Vannevar Bush, a much-acclaimed electrical engineer and the top science and technology advisor to Presidents Roosevelt and Truman (Bush 1946; Zachary 1997). In these essays, basic science was viewed as operating on an "endless frontier" on which progressively "better" (i.e., truer and more useful) knowledge would be discovered. Natural scientists and engineers (and to a lesser extent, social scientists) were viewed as pioneers on this frontier. The discoveries made by these pioneers would subsequently provide innumerable opportunities for practical applications. Through the practical applications that would be "spun-out" of basic science, universities would play an important role in national development.

The linear model also held that it was difficult to predict what basic research results would yield knowledge leading to practical applications and that whatever knowledge produced through basic research would be difficult for business firms to protect. Thus, though the potential benefits of basic science to society as a whole were clear, entrepreneurs from the private sector would invest less in basic science than the overall social return would justify. In

Figure 4.1
The Linear Model of Basic and Applied Research

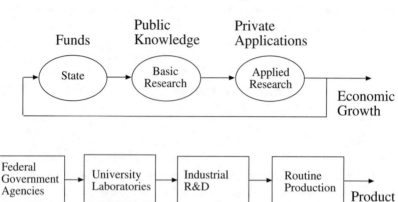

the vocabulary of neoclassical economics, basic research activities were the mechanism through which an important category of "public goods," foundational scientific knowledge, would be produced (Nelson 1977; North 1981; Callon 1994). Because the perceived risks for private sector research were too high (and the incentives too small), government funding of basic research in university and federal laboratories was justified as being in the national interest.

During the 1970s, the linear model of basic and applied research began to be challenged within U.S. science and technology policymaking circles (Shapley and Roy 1985). Among the events that spurred these challenges were the emergence of high-technology firms based in Japan, Germany, and other nations as formidable competitors to American industrial firms in both American and global markets and the new linkages forged between universities and industrial firms in the biotechnology industry. The emergence of foreign high-technology firms demonstrated the advantages of more predictable "spin-offs" from basic research into commercial applications and more reliably directing these results to American industrial firms (Porter 1990; Kogut 1993). The growing linkages between universities and private firms in the biotechnology industry provided a visible demonstration of the possibilities opened up by closer university/industry research relations (Kenney 1986; Teitelman 1989). Policy discussions began to focus on more than the need to support basic research as a foundation on which practical applications might be developed. In addition, policy discussions begun to focus on how basic research "assets" might be better deployed to foster a more collaborative mode of knowledge production out of which more practical applications might be spun out (Gibbons et al. 1994; Ostry and Nelson 1995).

In the late 1970s and early 1980s, these concerns spurred the passage of enabling legislation such as the Bayh-Dole Act of 1980 and the establishment of a variety of programs to encourage university-industry research collaboration at both federal and state levels. While these initiatives were being put into place, new models of research and development began to emerge in neo-Schumpeterian and organizational economics, organizational/occupational sociology, the sociology of science and technology, and management studies. These new models recognized that the development and subsequent transfer of basic research results into applied research and development was a much more complex process than the image of science as an endless frontier allowed for (Latour 1987; Utterback 1994; Rosenberg and Nelson 1993; McKenzie 1996).

In contrast to the image of scientific knowledge as an irresistible force for progress that would be naturally picked up by "rational" entrepreneurs operating in free markets, the tacit and local aspects of knowledge and the embeddedness of knowledge in specialized occupations and specialized instruments was acknowledged in these new discourses. In the words of Latour (1987), new knowledge and new inventions produced through the conduct of basic research did not simply "diffuse" out of universities and public-sector laboratories. Rather, they had to be "translated" from a basic research context into an applied commercial context (132–141). In the new models, for basic research results to be better translated into commercial arenas, faculty and universities needed to become more "entrepreneurial" (Brett, Gibson, and Smilor 1991; Etzkowitz 1997; Slaughter and Leslie 1997). To complement these changes in universities, industrial firms needed to improve their capacity to absorb and build upon the new knowledge spun out of basic research (Cohen and Levinthal 1990; Rosenberg 1990).

Figure 4.2 presents a process model of basic research in the "entrepreneurial" university, and figure 4.3 presents a process model of applied research in the "learning" business firm. Both models recognize the tacit and local nature of much knowledge and the iterative processes involved in research and development. Although these models taken together represent an improvement over the very simple linear model presented in figure 4.1, the structures that govern the movement of knowledge assets—people, technology, documents—remains masked. What is missing are all the administrative departments in both entrepreneurial universities and learning firms—and the public and private sector "brokering" organizations that mediate between universities and business firms—that are involved with regulating the flow of basic research results into commercial arenas.

Figure 4.4 presents a process model of the additional work for which these administrative units and brokering organizations are responsible. Rather than being directly involved with the work of knowledge production and deployment, these administrative units and brokering organizations work on tasks like

82

## Figure 4.2
### Basic Research Projects: A Process Model

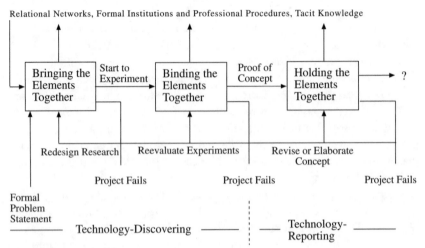

Basic Research Practices

Relational Networks, Formal Institutions and Professional Procedures, Tacit Knowledge

## Figure 4.3
### Applied Research Projects: A Process Model

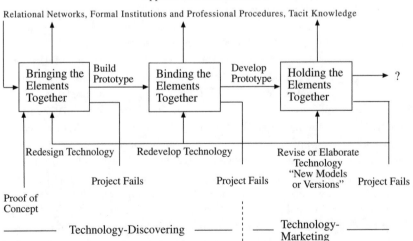

Applied Research Practices

Relational Networks, Formal Institutions and Professional Procedures, Tacit Knowledge

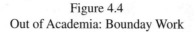

Figure 4.4
Out of Academia: Bounday Work

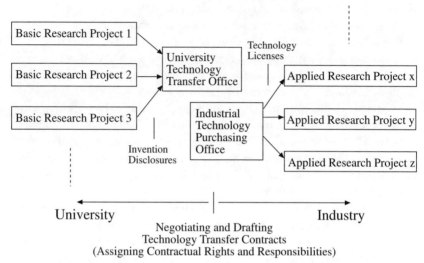

securing and managing intellectual property rights and negotiating and managing contracts through which exchanges involving knowledge assets are governed.
In addition to these formal duties, these administrative units and brokering organizations are caught up in trying to contain "boundary disputes" that revolve around
the academic values that continue to hold sway in entrepreneurial universities
and the commercial values that dominate learning firms in the private sector.
All of these tasks fit the general category of "articulation work" outlined by
Strauss and his students (Fujimura 1987; Strauss 1988). Articulation work is all
the extra work that goes on to insure that the various activities that make up
large-scale collective action stay coordinated. As will be demonstrated in the
following section, how these tasks are handled has a significant influence on
how knowledge flows "out of academia" and how discoveries and inventions
made on the "endless" frontier come to be exploited.

## HARNESSING PUBLIC CONGLOMERATES: TECHNOLOGY
## TRANSFER MANAGERS AND THEIR WORK

During the 1970s, at the same time that the linear model of research and development began to be challenged, researchers in organization theory began to
develop models of "organized anarchies" and "loosely coupled systems."[1] Universities were held to be the primary exemplars of these sorts of organization

(Cohen, March, and Olsen 1972; Weick 1976). In the "garbage-can" models associated with organized anarchy, the central administration of an organization is pictured as relatively weak in comparison to the different operational units of the organization, and organization-wide decision-making processes are difficult to coordinate and control (Cohen and March 1974). Relatedly, in models of organizations as loosely coupled systems, the different units that make up the organization (e.g., schools, colleges, and departments in a university) operate relatively autonomously on a day-to-day basis but are held together by some set of overarching common interests. Attempts to tighten the organization of loosely coupled systems such as universities are often viewed as reflecting ill-considered "rituals of rationality" that have a negative impact on organizational performance. The clear implication was that universities were better off maintaining a loosely coupled structure (Meyer and Rowan 1977, 1978; Stinchcombe 1990).

Curiously, over the same period a number of researchers in the strategic management of business firms suggested that business firms constructed as loosely coupled systems would, in comparison to more tightly coupled organizations, perform poorly. However, rather than talking in terms of organized anarchies or loosely coupled systems, these researchers talked about diversification strategies that resulted in a conglomerate structure (e.g., Rumelt 1974; Porter 1985; Barney and Zajac 1994). Of particular importance for these researchers are the problems that conglomerate firms face in promoting entrepreneurial activity and managing innovation across unrelated lines of business (Pisano 1994). Arguably, the term *entrepreneurial conglomerate* when applied to a business firm is an oxymoron. The typical advice to a firm with a conglomerate structure that wants to innovate is, ceterus paribus, to divest some of its lower-performing assets and use the resulting funds to promote innovation in the consolidated firm. In the terms of the previous paragraph, the advice would be to make management less anarchic and the firm more tightly coupled.

Taken literally, the perspective of the strategic management of private business firms implies that universities, as organized anarchies/loosely coupled systems, are the worst sort of conglomerates and deserve to be broken up. Yet recently some researchers in strategic management (e.g., Argyres and Liebeskind 1998) have argued that universities engaged in basic research do not have the same function as entrepreneurial firms doing applied research. Stated differently, universities are viewed as public rather than private conglomerates. In this view, the conglomerate structure of universities is justified because basic scientific research is an important public good. Relatedly, the sort of performance that various external stakeholders demand of an entrepreneurial university is likely to be quite different than the performance that stockholders demand of entrepreneurial firms. Rather than focusing on profit maximization, in principle entrepreneurial universities are supposed to maximize the socially beneficial effects of (loosely coupled) university research and teaching.

One important socially beneficial effect involves spinning university-generated inventions and know-how into practical applications. The demand for the more effective spin-off of basic research results into applied research and development while protecting the basic research process involves the flexible creation and adaptation of governance structures. The creation and adaptation of governance structures is the core of the articulation work performed by specialized university technology transfer offices and the professional technology transfer managers that staff these offices (Stinchcombe 1990).

In their own words, technology transfer managers are responsible for "arranging deals." Arranging deals involves constructing and monitoring both formal and informal understandings among the various parties to the deal so that a complex exchange relationship can be performed satisfactorily. Technology transfer managers, as intermediaries between all the parties, are responsible for making sure that the collection of formal and informal understandings that form the basis for the deal remain as mutually consistent as possible over the course of performing a deal. This articulation work helps ensure that the mutual expectations of all the parties involved in performing a deal remain realistic in the face of various contingencies that arise as a contract is being performed. Less formally, this work helps ensure that potential deals become done deals and that done deals remain "doable."

Figure 4.5 provides a simplified picture of what arranging the formal understandings (typically embodied in a written contract) for technology transfer relationships entails. A signed contract between a university and an industrial firm provides a formal and authoritative representation of the "meeting of the minds" that all the parties to a relationship have knowingly arrived at about how a particular set of mutual obligations should be structured and performed. A signed contract between a university and business firm is typically supplemented by a number of ancillary agreements. In an industrial firm, these ancillary agreements might be "work-for-hire" employment contracts for research and development employees that ensure that the firm rather than the employee controls any property rights that result from contracted research work. In a university, these ancillary agreements might include written memos of understanding that set out the mutual responsibilities of the central administration and the departments and researchers involved in performing the actual work involved in a collaborative research relationship. These formal understandings constitute the "contract space" in which ongoing exchange relationships will be performed over time.

Figure 4.5
Arranging Deals as Articulation Work

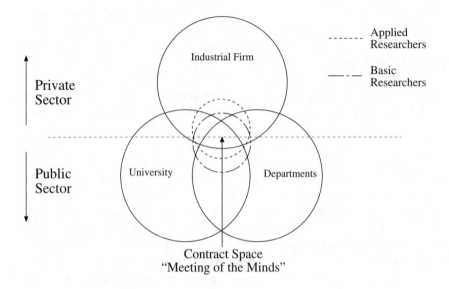

Contract Space
"Meeting of the Minds"

In addition, as a number of studies have shown, formal contracts and similar formal agreements are almost always supplemented by a variety of more
informal understandings that help the contracting parties carry out the pledges
that they have made in the formal contract (Macauley 1963; Clegg 1975; Uzzi
1996). This task is complicated by the number of parties that must be tied
together in the construction of a workable deal. As shown in figure 4.5, in the
simplest technology transfer deal, the informal expectations of at least four parties—researchers, academic departments, the university, and business firm as
corporate entities—must be considered by technology transfer managers in the
course of arranging deals.[2] Although technology transfer managers are formally
responsible for negotiating with business firms on behalf of the university, they
inevitably become entangled in negotiations between university researchers
and their departments, university researchers and the staff of the business firm,
and academic departments and the university central administration. How these
various negotiations are reflected in the formal and informal understandings
that constitute the deal has a great deal to do with the ways in which problems
that arise in the course of performing the deal can be resolved.[3]

Importantly, the work of arranging deals involves a process that extends
roughly from the time a university-generated invention is first disclosed to the

Figure 4.6
Arranging Deals: A Process Model

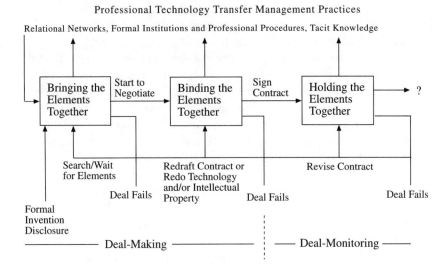

Professional Technology Transfer Management Practices

Relational Networks, Formal Institutions and Professional Procedures, Tacit Knowledge

technology transfer office and a new file opened until the time that the file associated with the invention disclosure is formally inactivated. Figure 4.6 presents
a general model of the stages and processes involved in arranging deals. The
stage labeled "bringing the elements together" roughly corresponds to a search,
analysis, and choice stage (March 1988). The stage labeled "binding the elements
together" roughly corresponds to a negotiation/structuration stage (Strauss 1978;
Giddens 1984). The stage labeled "holding the elements together" roughly corresponds to a performance phase in which monitoring and enforcement are
important issues (Williamson 1985).

However, this correspondence is only approximate because the model
emphasizes the interdependence of all three stages and the mutability of ongoing deals as they progress over time. The "bringing the elements together"
stage ends not in a choice but in the start of a negotiation process. The "binding the elements together" stage emphasizes the amount of negotiating work
involved in constructing specific contractual arrangements and the larger legal
and policy structures by which these negotiations are constrained and the effect
of the other two stages on the work that goes on in this stage. The "holding the
elements together" stage emphasizes that negotiations and learning continue to
be important elements as a deal is performed and that monitoring and enforcement arrangements are often adjusted in the face of unresolved misunderstandings and unexpected contingencies. Furthermore, how well these monitoring

and enforcement mechanisms work often reflects the quality of the deal-making processes out of which these structures were forged.

This model also helps to clarify some additional problems with the linear model of basic and applied research that are not well addressed in the recent research that emphasizes the importance of effectively communicating tacit knowledge during the research and development process. In particular, the model indicates the importance of the work that is done in binding the parties to a technology transfer relationship together through a formal contract that is signed by authorized representatives of a university and an industrial firm. Encouraging faculty to be more entrepreneurial in pushing their research into practical appli-cations would seem to help in the search for industrial firms to partner with and in the faculty's ability to monitor the behavior of the industrial partner after a contract has been signed. Similarly, encouraging business firms to retain a capac-ity for quickly absorbing and responding to the new opportunities opened up by basic research are important elements in finding and exploiting new and promising technologies. Nevertheless, once a basic researcher and an applied researcher have found each other, they must still be bound together in such a way that the traditional boundaries between university and industry are respected. This is the work in which specialized technology transfer offices and professional technology transfer managers take the most interest.

In the course of doing this "binding" work, technology transfer managers are constantly made aware of the different ways in which the worlds of basic and applied research are governed. In the world of basic research, specialized academic disciplines and university departments are typically the locus of work, evaluation, and reward. Disciplines are concerned with building and maintain-ing certified knowledge in their particular theoretical and empirical domains. Disciplinary evaluation is typically organized around publication in recognized journals or academic presses, on the open sharing of information among mem-bers of the discipline, and on the open transmission of knowledge to the broader society (Crane 1972). Individual departments are important operational units within disciplines. Departments are particularly concerned with how the research produced by the department as a unit will be rated within the disci-pline. The manner in which departments are governed helps to maintain disci-plinary standards at a local level. This system roughly reflects what Gibbons and his co-authors (1994) have termed a Mode 1 system of knowledge production. From this perspective, basic research is fundamentally a "public" activity and basic research results are "public goods" (Callon 1994).

In the world of applied research, on the other hand, organized professions and business firms are typically the locus of work, evaluation, and reward. Professions are concerned with maintaining and monitoring standards over the particular collection of practices in their professional domain and maintaining the position of their profession in a system of professions (Abbott 1988). Business firms, in pursuing research and development projects, draw on the professional skills of experts in a number of different technical and management areas to respond to perceived market opportunities and competitive threats (Brown and Duguid 1991; Dougherty 1996). The interdisciplinary cross-functional teams of experts involved in applied research roughly reflect what Gibbons and his co-authors have termed a Mode 2 system of knowledge production. However, unlike university departments, business firms operating within the private sector of capitalist and mixed economies have a particular concern with maintaining some form of control over their "knowledge assets" (Teece 1986). From this perspective, applied research is fundamentally a "private" activity and applied research results are "private goods" (Nelson 1989).

Managing the translation of a university-generated technology from a "public" to a "private" activity represents a great challenge (Webster and Packer 1997). Not only does the concrete knowledge or know-how about a subject technology have to be translated from the university laboratory to the industrial laboratory, but the contradictory world views of basic and applied research must be reconciled in such a way that both public and private sectors are respected. This work is made even more complicated by the volume and sheer heterogeneity of the inventions that are disclosed to a technology transfer office and the potential applications to which these inventions might lend themselves.[4] Under these conditions, it is difficult (and very likely ill considered) for individual managers or for the office as a whole to standardize on particular deal-making procedures and practices. Rather, the individual managers tended to develop work routines that allowed them to improvise as a subject invention progressed (hopefully) from being "something" summarized in a paragraph on an invention disclosure to being a more well-defined object of exchange in a binding contract (Orlikowski 1996; Weick 1998). Similarly, general office policies and procedures, though represented formally in policy statements and official department literature, are in practice flexible enough to allow for sensible responses to new situations.

In sum, arranging deals is an "organic" rather than a "mechanistic" process (Burns and Stalker 1961). To translate basic research results from public conglomerates such as universities into applied research projects in private business firms on a routine basis requires that a great deal of attention be paid to the terms under which these translations are made. Professional technology transfer managers work to insure that these terms are well articulated and that the activities involved in performing the deal remain well matched over time. As illustrated by the three stories about software technology transfer in the following section, this articulation work involves synthesis as well as analysis, art as well as science.

ARRANGING SOFTWARE DEALS

## 1. The Emergence of Software Technology Transfer Practices in a University Technology Transfer Office

The Office of Technology Transfer (OTT) was established in the early 1980s to handle several administrative tasks. These included receiving and evaluating invention disclosures, reporting the disclosure of patentable inventions that had been developed with federal funding to the federal government, passing these inventions to a sister organization, the Washington Research Foundation (WRF), for licensing, and distributing any revenues generated by licensing to the proper recipients within the university. During the 1980s, one particularly lucrative invention (in biotechnology/health care) was patented and licensed. This single invention returned yearly revenues in the millions of dollars to the university and, arguably, became the model of a desirable technology transfer deal. WRF, rather than OTT, managed this particular license.

Over the course of the 1980s, as more inventions were disclosed to OTT, WRF became more selective about the disclosed technologies that it was willing to invest time and money in managing. It took particular interest in potential "big-hit" inventions out of biotechnology. However, some researchers who had disclosed inventions and the technology transfer managers at OTT responsible for evaluating these inventions felt that the WRF rejected some inventions that had promise. To pursue deals on these "rejected but still promising" technologies, the technology transfer managers employed by OTT began to develop deal-making skills that supplemented those of the managers at WRF. The managers at OTT focused particular attention on those promising technologies that had profiles different from those favored by WRF.

In the late 1980s, the OTT director of the office became aware that an increasing amount of software was being disclosed to the office. This software was diverse, ranging from PC-based digital media applications for classroom teaching to systems software, such as operating systems or compilers. Very little of this software was "patentable"[5] but some of the software clearly had commercial potential, the software developers, and their departments and the university administration were often quite interested in investigating these possibilities. However, managing software technologies was not easy for the office, because software did not fit well with the standard practices of the office that were tailored for handling "patentable inventions" such as those that arose out engineering and biotechnology.

Because of these problems, a new position—software technology manager—was established in the late 1980s. This manager was given the mandate to develop systems and practices for handling these sorts of technologies either by himself or in tandem with other managers in the office.[6] The software technology manager was by some measures very successful. In particular, the contribution of software to the total revenue produced from the commercialization of university-generated "inventions" greatly increased both in absolute terms and as a percentage of total revenue. However, both the manager's position in the office and the status and distinctiveness of practices for handling software technologies in the wider community of specialized university technology transfer offices remained precarious for a number of years, including the first couple of years of the field study.[7]

By the end of the field study, however, the position of the software technology manager had changed dramatically. He had received a substantial raise and was given greater discretionary authority for handling his work. A second software technology management position was created to help with the increasing workload. Six months after the end of the field project, the technology transfer office was reorganized and the software technologies manager became the head of a "software unit" that had its own separate office on campus located close to the computer science and engineering departments. This software "unit" hired additional clerical staff to help with the work. The change in circumstances was rather remarkable.

Although the software technologies manager was clearly a talented individual who had established a powerful set of work routines, it was equally clear that these changed circumstances could not be attributed to his efforts alone. At a local level, many of the routines he developed helped him to do his own daily work; he had worked hard to develop close working relationships with inventors, lawyers and others who he interacted with on a daily basis. In particular, his work routines were well articulated with those of others whom he had to interact with on a routine basis. This articulation required a great deal of ongoing learning and mutual negotiation.

At a somewhat higher level, in line with efforts of its "peer institutions," the university was trying to upgrade its technology transfer capabilities during the period of the field research. To this end, the office was reorganized and a new director was hired. In contrast to the old director, the new director had a Ph.D. in biotechnology, had managed a laboratory, and brought a very different management style to the office. The new director was committed to earning the university greater recognition in technology transfer, recognized the importance of software revenue to office performance, and endorsed the efforts of the software technology transfer manager. However, the new director's appreciation of the importance of software was also connected to the emergence of the Internet and the centrality of university-generated systems software for the Internet as an important commercial arena. Although Internet-related software was not the only (or even the typical) sort of software that the software technologies manager handled, his successes in this arena helped to legitimate the often quite different set of practices that he had developed to handle different sorts of software. This was reflected in the growing reputation of the university within the technology transfer community as being a "leader" in software transfer practice.

## 2. Establishing Office Strategies and Practices for the Management of Software Technologies

During my stay at the field site, new markets began to emerge in both systems software and applications software that had a significant impact both at the University of Washington and on the strategy and structure of the specialized technology transfer office. In applications software, special-purpose multimedia applications that translated university expertise into a new electronic medium began to be disclosed to the office from a wide variety of departments across campus. Although few of these applications held the promise of being a large individual source of revenue, in the aggregate they represented a steady source of income. Furthermore, these products promised to enhance the university's reputation as a leading educational institution as well as a leading research institution. In systems software developed within the computer science department and by staff members of the university computer and communications staff, advances in areas such as the Internet and advanced computer graphics showed commercial promise.

As has been detailed in Kaghan and Barnett (1997), the software manager and his clients found that "doing software deals" was quite different from the sort of deals that the office expected. The impact of this "software-specific" knowledge was evident in differences between the software manager and various other professionals involved in technology transfer operations about the proper way to manage software technology generally and about possible alternatives for managing specific software technologies. Of particular interest was the way in which the language used by managers in the office to discuss intellectual property issues related to software transfer reflected these differences.

In particular, managers other than the software manager often used the term *intellectual property* as a synonym for *patent* when the primary formal property rights in software were copyright and trademark. In the software manager's eyes, equating intellectual property rights with patent rights in this way indicated a myopia that prevented these other professionals from appreciating the wide variety of opportunities for software transfer that were opening up. By habitually confusing the opportunities afforded by patent rights with copyright or trademark, the other managers signaled to the software manager that they might make serious errors if they tried to manage software technology.

If the daily operations of the office just involved the software manager listening to other managers use the words *intellectual property* as a synonym for *patent* when talking about software, he would likely have adapted to it. However, this patent-bias was embedded in office practices in other, more troubling ways. For example, the language of the standard boilerplate licensing agreement rested on a default assumption that a patentable technology was being licensed. Most (if not all) of the clauses that made up the contract are importantly affected by the underlying property rights attached to the technology being transferred. Thus, the standard boilerplate contract needed to be almost completely rewritten in the course of drafting software licenses. The software manager had, in fact, written his own standard software license that he could reuse. However, if other managers attempted to use his boilerplate software license as a model for software licenses that they were attempting to write, they often found themselves entangled in misunderstandings caused by their need to translate any concept of property rights into patent rights.

These sorts of situations were particularly difficult when the software manager had to co-manage with another manager a technology that contained both hardware and software elements. Almost invariably, the copyright and trademark protection appropriate to the software portion of the technology became invisible and was thus not considered in the process of arranging deals. Similarly, the software manager often had to submit particular agreements to others in the office for signature approval. Predictably, the people reviewing the documents would miss certain nuances where copyright or trademark were involved or make objections based on patent-biased understandings.

In a sense, this discussion points to the barriers that were erected to effectively arranging software deals because of the patent-biased dialect of technology transfer legalese that dominated the office (as well as many important stakeholders—senior university administrators, researchers, industrial partners—outside the office). In a situation in which software technology was not a significant element of the technology portfolio, this problem would have been less troublesome. However, software was a significant and growing part of the office's business. The difficulty that the software manager had in getting the copyright and trademark dialects understood (or even recognized) showed in a rather negative way how the culture of technology transfer and the practices of technology transfer management were intertwined. As software deals came to be recognized as important elements of the technology transfer office's business (and after the office was reorganized to reflect the importance of software technology transfer), the language used to discuss intellectual property rights of the office was gradually modified.

## 3. Negotiating and Drafting Software Licensing Agreements

Stinchcombe (1990, 195) defines an "ideal-type" contract to include "all market transactions in which one firm or person makes an offer and another
accepts." He argues in a related way that transaction cost economics and other
views of contract grounded in economics predict that whenever it is difficult
to specify the required contractual performance in advance, hierarchy is prefer-
able to market coordination through contracts as a governance mechanism.
However, he pointedly notes that there are numerous transactional situations in
which these expectations are not met, and he explores some of the ways in
which contracts that depart from the ideal type can be structured so that
uncertainties about performance can be addressed. However, in his subsequent
discussion he does little to explore how such non-ideal-type contracts come to
be written in practice. In particular, he does not address how "negotiating
costs" (ex-ante) and "managing costs" (ex-post) are interrelated through the
negotiating and drafting process.

Observing the work that professional technology transfer managers perform confirms Stinchcombe's observation that contractual relations are used in
exchange situations in which performance is difficult to anticipate. In addition,
these observations help to illuminate the manner in which these non-ideal
contracts are actually negotiated, drafted, and signed. In particular, the work
that professional technology transfer managers perform indicates that the inter-
actional process envisioned for an "offer" and "acceptance" between a university and an industrial firm is much more extended and much less determined
than traditional views of contracting allow. But it is precisely this extended and
flexible period in which the elements involved in a technology transfer deal are
bound together which make contracts a viable alternative to bureaucratic hierarchies. Furthermore, the work of professional technology transfer managers
demonstrates that the deep understanding and skilled use of technology transfer language and practice is an essential element of the negotiating and drafting
process.

In Stinchcombe's idealized picture of the offer/acceptance process, the
process is reduced to a single event. In contrast, the offer/acceptance process in
the majority of deals observed during the field study involved a whole range of
activities that extended over the phase labeled "binding the elements together"
in the previous section. This phase of "deal-making" begins when the university and a business firm first seriously explore the possibility of setting up a
licensing or some other contractual arrangement for a university-generated
technology. Although different managers (and different firms) had somewhat
different practices with regard to this initial "feeling out" process, these practices can be roughly divided into two ideal types. In the first sort of practice,

the manager and his counterpart in a business firm would exchange "blue-prints" (Weick 1979, 47). In the other sort of practice, the manager and his counterpart would exchange "recipes."

In the first sort of practice, managers and their counterparts in firms would begin working from the relevant template agreement that their respective organizations would customarily use in this sort of transaction. They would then work to find a recipe that allowed the two blueprints to be merged into a concrete contract. In the second sort of practice, in contrast, managers and their counterparts preferred to begin by exchanging outlines of the key elements of the transaction that were used to reach agreement in principle before serious drafting began.[8] The first sort of practice typically indicated a desire by the university or the firm to maintain a standard and well-understood type of contractual arrangement. The second sort of practice typically indicated a somewhat more accommodating stance on the part of the university or industrial firm about the sorts of business relationships with which they were willing to be involved given the key elements of a deal.

In practice, the situation was rarely as clean-cut as described in the previous paragraph because neither the technology transfer manager nor his counterpart in an industrial firm worked in a vacuum. For example, managers who were inclined to work from "heads of agreement" might be constrained by an organizational policy that encouraged conformance to a template. Conversely, managers might be forced to abandon templates and "backtrack" to an exchange of "heads of agreement" if the respective templates could not be fit together in a particular situation and the university and the industrial firm still desired a business relationship. Whatever the exact sequence of events, technology transfer managers (whether they preferred to start with blueprints or recipes) had to come to understand the development and marketing concerns and the basic business models of the industrial firm and how these concerns might be fit with university concerns and business models in the context of a particular deal. Technology transfer managers and their counterparts then had to translate these understandings into formal legal language and sell this contractual language to the faculty members and senior administrators. Technology transfer managers' industrial counterparts often talked about doing similar sorts of selling work within their respective organizations. But within industry, the job often revolved around convincing decision makers that a university-generated technology was worth licensing despite the additional development, marketing, and contractual complications.

For example, attempts (actual or perceived) on the part of an industrial firm to set up nonstandard relations with the university or on the part of the university to set up nonstandard relations with the industrial firm created difficulties. These difficulties required a great deal more effort on the part of technology transfer managers and their industrial counterparts in negotiating and drafting a substantive legal agreement. This work was particularly difficult when the university departments responsible for developing the subject technology and industrial firms attempting to license the technology had little experience with university-industry technology transfer and the technology being transferred was not the typical sort of technology that the technology transfer office handled. In these cases, the academic unit involved was not familiar with the standard ways in which university central administration (as represented by the technology transfer office) might expect to arrange university-industry relations (as reflected in university template contracts). Furthermore, the technology transfer office as an administrative unit or the technology transfer manager responsible for negotiating and drafting deals often needed time to get a full sense of the potential of the technology and the concerns of the inventor and his or her department. Conversely, the industrial firm, which looked on the university as a single coherent business unit, was unaccustomed to dealing either with university administrations or with individual university departments. In situations like this, university and industry business models were rarely well matched and contracts had to be hammered out clause by clause (and sometimes term by term) until a satisfactory document that could be signed by authorized representatives of the university and the industrial firm was produced.

These sorts of issues were exacerbated when either the university or the industrial firm customarily offered a standard template agreement (i.e., a blueprint) as a basis for negotiating and drafting a concrete contract. In this case, standard business models that were a poor fit for the emerging deal were embedded in the formal structure and the legal language of a template contract. This embedding created a variety of problems. Technology transfer deals were not routine activities for the majority of senior administrators and faculty inventors, and nonstandard business arrangements simply exacerbated this feeling of unfamiliarity. This general feeling was heightened because the nonstandard business arrangements could not be easily read out of the legal language in which the template contract was composed. Except in rare instances, they had limited insight into the full legal import of the documents. Furthermore, though they had some feel for the basic business model that they as administrators or faculty members favored, they experienced difficulty in expressing these desires in a legally persuasive form. In an environment where the differences between business models for the development and transfer of software products and business models for the development and transfer of patentable inventions were not well understood, these problems were more pronounced. For the software technology transfer manager directly involved in negotiating and drafting an actual contract, this work presented a severe test of his skills.

To deal with these sorts of problems, the software technology transfer manager had to be able to read the industrial business model out of the boiler-plate contract and translate it into language that was understandable to both senior administrators and faculty inventors. In doing this, the technology transfer manager had to convey some sense of the marketing and development concerns of the industrial firm that might present real obstacles to commercial development while not totally devaluing or demeaning the contribution of the university. At the same time, the software technology transfer manager had to convey to his industrial counterparts the genuine nature of university concerns. Sometimes deals failed to materialize because the goals or skills of the university and the industrial firm were not well matched. Nevertheless, there was the ever-present danger that potential deals would be missed because of miscommunication centered around an inflexible reliance either on the part of managers themselves or on the part of others involved in the deal-making process on mechanisms such as standard template contracts. In particular, conflicting expectations on the part of the university and the industrial firm had to be recognized and then had to be painstakingly reconciled as universities and their industrial partners struggled to articulate their desires into mutually acceptable contractual language.

In doing this work, the software technology transfer manager (and his industrial counterparts) had to be able to conceptualize jointly the progressive disentanglement of the technology from its close university connections and its progressive entanglement in an industrial firm or a collection of industrial firms. Once this common conceptualization was reached, they had to be able to memorialize the basic structure of that development process in a formal contract and to explain that structure in a language that made sense to their faculty and administrative clients. An inability to perform these various translations was sufficient to break up promising potential deals.

## CONCLUSION

Many authors have detailed the multiple contingencies involved in scientific research and technological development. For example, Hughes (1992) talks about the development of large technological systems in terms of a battlefield metaphor where system builders meet and overcome "critical salients" at various points in time. Though he acknowledges that system builders like Edison or Ford do not work alone, he does relatively little to fill in the tactical details that they employed in particular skirmishes. Utterback (1994) gives some sense of what these "tactical skirmishes" might be like by comparing technology development to the children's game "Chutes and Ladders," where the path to successful development is lined with unpredictable ladders that speed development and unpredictable chutes that impede development. But as Weick (1990) notes, these technological chutes and ladders, as they are confronted and experienced, are much more equivocal than is commonly supposed. Climbing ladders and avoiding chutes often requires that all the people involved in technology development be able to quickly and jointly define a situation and adapt the sociotechnical system in which both developers and technology are embedded in "appropriate" but ill-defined ways.

As this chapter has tried to make evident, the translation of new technological systems from conception to prototype to working system depends on much more than just the development and production activities performed in laboratories and on factory floors. People working in offices are as involved with the translation of technologies as people working in laboratories or on factory floors. They simply deal with technologies in different ways. For example, institutional and evolutionary economists (e.g., Nelson 1990; North 1990) have pointed out the importance of formal economic institutions, such as intellectual property rights and formal contracts, to supporting collective economic activity. Economic sociologists, following Weber, (e.g., DiMaggio 1994; Hamilton 1994) have stressed the importance of a widely shared cultural ethos about topics such as public and private goods to economic activity. However, relatively little attention has been paid to the equivocality surrounding the formal institutions that the institutional economists discuss or the work involved with managing this equivocality.

Successful university-industry technology transfer operations rely not on total "consensus" among "rational" individuals but rather on a "meeting of the minds" among "pragmatic" but culturally bound individuals and groups. This "meeting of the minds" is symbolically represented by a contract and is recognized as binding by courts of law. But this "meeting of the minds" is quite narrow and specific. Only the very most important points about a technology and the relationships required for the ongoing development and transfer of the technology are "hardened" in contractual language. This "hardened" language is the product of intense negotiation. Managing these contracts requires sensitivity to the context in which this "hardened" language was produced. The importance of the work of technology transfer managers rests precisely on the fragility of the terms on which a "meeting of the minds" is constructed. The activities involved in technology development and technology transfer place these understandings under extreme stress.

Similarly, it is crucial to recognize the contingent nature of any routine practice for the transfer of university-generated inventions into commercial arenas. Routines are continually tested in interaction, and the contingencies through which these routines are tested may arise from either the local or the global environment. What was "best practice" in one set of circumstances may turn out to be inapplicable in other circumstances. On occasion, new "best practices" must be synthesized. In short, collective responses to these sorts of contingencies inevitably demand some form of struggle and negotiation. Among the most important struggles are struggles over boundaries (Zerubavel 1991). In resolving the boundary disputes that arise on the borders of basic and applied research, professional technology transfer managers play an important and often unrecognized part in harnessing the "public conglomerates" that entrepreneurial universities are.

## NOTES

1. The discussion in this section is based on evidence collected during a participant-observer field study that was conducted at the University of Washington Office of Technology Transfer from May 1993 to June 1996.

2. As researchers working in the Actor Network Theory (ANT) tradition have emphasized, "nonhumans" such as the subject technology in a technology transfer deal might also be considered "parties" to the deal. From an ANT perspective, the resistance of the technology to being developed along particular lines has a great influence on the direction that a deal may take (Callon 1986; Latour 1993, 1996; Law 1994). In the context of modern Western business law, "nonhumans" are not viewed as genuine contracting partners but are rather "represented" in particular ways by the various human parties to the deal. However, as Strathern (1999) has pointed out, this assignment of contracting rights to humans and nonhumans is peculiar to the modern West and not to all cultures.

3. The presence of additional parties, such as other universities, government agencies, multiple departments, or multiple business firms, or a lack of a previous technology transfer relationship, as is often the case, requires that even more parties be involved in some fashion in the binding stage.

4. When the field research on which this chapter is based commenced, the technology transfer office had over a thousand active files on inventions disclosed to the office. Over the three years during which the field research was conducted, nearly 400 new inventions were disclosed to the technology transfer office. The majority of these invention disclosures had been made from the natural sciences, engineering and computer science, the health sciences, and researchers in a variety of other departments (e.g., dentistry, ethnomusicology, the language laboratory) had made disclosures as well. Even within the College of Engineering or the Health Sciences School, the individual research projects from which these invention disclosures were generated were, at best, very loosely related. When the field study commenced, there were four full-time technology transfer managers and an associate director of the office. When the field study ended, there were six full-time managers and a full-time director of the office.

5. Copyright and trademark were typically the applicable property rights in software. Software patents were rare and were a controversial area of intellectual property law. The university at which the research was conducted held no software patents.

6. The software technologies manager kept meticulous notes of all his interactions on coded green sheets of paper. One of his major problems was trying to file these records. When the field research project commenced, this manager's office stood out because of the large number of piles of seemingly loose sheets of paper stacked on desks, shelves, and on the floor. One of the major work projects assigned to the field researcher in his capacity as a technology transfer intern was to construct a filing system to manage these papers and to file these papers away. In his capacity as an investigator, the researcher spent many hours reading and analyzing these green sheets of paper to get a better sense of the development of this particular manager's work routines.

7. There is not sufficient room in the paper to demonstrate this assertion. However, the researcher's fieldnotes contain numerous indications of this situation. In the case of the software technologies manager himself, he remained the lowest-paid manager for a number of years despite the revenue that software brought in, had numerous disputes about the performance of his duties with his immediate superiors, and visited the university ombudsman about these problems. Within the population of research universities, the difficulties of universities in establishing formal policies for handling software was evidenced by an investigation of these policies that the researcher conducted on behalf of the software technologies manager, by numerous articles written for professional technology transfer managers and in the correspondence on Techno-L, an Internet listserver on which professional technology transfer managers, communicated with each other about professional issues.

8. These outlines were called "heads of agreement" in technology transfer parlance.

# 5

# Science as a Vocation in the 1990s: The Changing Organizational Culture of Academic Science

## EDWARD J. HACKETT

What are the prospects of a graduate student who is resolved to dedicate himself (or herself) professionally to science in university life?

Max Weber (1918, 129)

### FOREWORD

"Science as a Vocation in the 1990s" was written in the late 1980s with the intention of using insights from a series of intensive interviews with scientists to discuss the circumstances and prospects of scientific work. If it is foolhardy to write such a paper, it is truly humbling to review it ten years later. The paper was written with three main purposes in mind. First, it sought to offer a systematic account of the circumstances of scientific work, circumstances that differed sharply from recent experience and conventional understanding. Second, it sought to use those observations as the foundation for speculating about the future of work in academic science. Finally, the essay tried to show the continuities between contemporary observations of academic science with those offered, some seventy years earlier, by Max Weber and Thorstein Veblen.

I have resisted the temptation to revise the essay to accommodate changes in knowledge or circumstances, instead limiting editorial changes to rewriting a few awkward passages and repairing ambiguities. An afterword summarizes some recent work that develops themes raised in the essay.

⟫•◦•⟪

It is of course hubris to invoke Max Weber's classic work, "Science as a Vocation," in the title of this essay. But it is unavoidable, as I will be concerned with much the same question: how might emerging and potential changes in the culture of academic science affect the work and careers of young scientists?

In his lecture, Weber analyzed the academic prospects of young scholars by first considering the external conditions that govern scientific careers in the university and then examining the internal meanings of science for the scientist. In that spirit this essay will analyze changes in the structure and values of academic science that have consequences for post-baccalaureate education in the United States. Specifically, I outline a framework for thinking about value change in academic science that views culture as a set of axes of variation and considers academic science as an organized activity shaped by general organizational forces, particularly the quest for resources and legitimacy. I then discuss, extend, and illustrate aspects of this perspective on academic science with interview material from a study of the consequences of federal funding for academic research (Hackett 1987). Finally, I propose some major underlying dimensions of value change and conflict in academic science.

## STRAINS ON POST-BACCALAUREATE EDUCATION IN THE UNITED STATES

Critical examinations of higher education in the United States have not spared the post-baccalaureate years of graduate study and postdoctoral work. In fact, such work preceded recent colorful tirades about the shortcomings of undergraduate education. A host of issues has been raised in volumes that assess "The State of Graduate Education" (Smith 1985), examine "Postdoctoral Appointments and Disappointments" (National Research Council 1981), and review "University Finances" (General Accounting Office 1986), to name just three. These commentaries and the analyses they contain are concerned with the reduction in resources for graduate study, declines in the level and quality of scientific instrumentation, increases in the proportion of foreign-born graduate students in U.S. universities, difficulties in the postdoctoral years, supply/demand projections for scientific person-power, and the like. Certainly these are important issues, and the empirical work and logical analysis underlying them are exemplary. But their analyses are not anchored in a model of the culture and organizational context of academic science. This essay will suggest some basic elements for building such a framework.

## THE CULTURE OF ACADEMIC SCIENCE:
## THE TRADITIONAL PERSPECTIVE AND A NEW VIEW

> Students looking at modern societies are usually less impressed by the congruities than by the incongruities in what they see. They are sensitive to the fact that human life is frequently a thing of ambivalence and tension, of contradictions and conflict; and these are exactly the things one hears the least about in most social science descriptions of culture.
>
> Erikson (1976, 80)

In social science, "culture" is a set of guiding values, beliefs, principles, rules, and material objects that shape human behavior. In this, the simplest of its meanings, culture is treated as a static (or slow-changing) complex which reflects the main currents of a society. Therefore, one studies culture in this sense by cataloging the fundamental properties and analyzing their consequences for behavior.

This has been done well for science, treated as a social institution, first in Merton's (1973, 267–278) classic formulation of the norms of science (communalism, universalism, disinterestedness, and organized skepticism) and in subsequent refinements and extensions (Barber 1952; Hagstrom 1964; Storer 1966). In contrast to these writings, Mitroff (1974a,b), pursuing Merton's (1973, 383–418; 1976) notion of sociological ambivalence, proposed that scientists are guided by both norms and counternorms, which are in tension. Thus the norm of communality, which enjoins a scientist to make public his or her work because it should become the shared property of the scientific community, is in tension with a norm that treats scientific results as private property, at least until one has derived sufficient career benefit from them. Other norms have corresponding counternorms, creating ambivalence in scientists who must reconcile these inconsistent principles of conduct. This is familiar terrain. The point to retain is that the ambivalence of scientists departs from the idea of a culture as certain tendency—an unequivocal blueprint for behavior—and introduces value tensions that blur individuals' judgment.

In *Everything in Its Path* (Erikson 1976), a book about technological transformation and disaster in an Appalachian community, Kai Erikson uses the motifs of individual ambivalence and community differentiation to explain cultural change. Culture, he argued, is best considered not as a central tendency but as a "moral space within which people live" (81), crosscut by several axes of variation, the poles of which are anchored by contrasting values.[1] At any point in time a society may be located along these dimensions, and external perturbations move the society in predictable directions along the axes. Culture, then, resembles a gravitational field in which the axes define potentials for movement. Historical events that disturb the society do not "create" new values and ethics out of whole cloth, nor do they necessarily pose novel value conflicts, but they instead alter the balance between pre-existing polar opposites.

Erikson's ideas about culture and cultural change suggest a new context for the work of Merton and Mitroff. In this view, Merton's original norms may reflect one aspect of science, perhaps the dominant one in the social forms of science he observed in his historical research, or perhaps the dominant form at the time of his writing. In either case the norms of science are historically situated accommodations to a particular set of circumstances, not universal principles that transcend time and locale.[2] Thus we might think of norms and counternorms as pairs of latent principles, always in tension. As circumstances change, however, one member of the pair is brought into prominence. In Mitroff's work, the competitive and politically charged atmosphere of space science research in the early 1970s brought the poles he labeled "counternorms" to the fore. Just as Appalachia's transformation was brought about in part by a changing resource relationship between mountain communities and the larger society (first through lumbering, then coal mining) and in part by a changing cultural relationship (facilitated by improvements in communication and transformation), so too is academic science reshaped by changes in its cultural and resource relationship with society.

Framed in these terms, the problem of understanding change in the culture of academic science becomes at once more tractable and more complex. It is more tractable because we now have a framework and language for interpreting contradictions in academic science, rather than dismissing them as measurement errors, irrational behaviors, or temporary turmoil. But the problem is also more complex because we must go beyond simply listing the norms of academic science and instead identify the dimensions that contain the norms and characterize the social forces which compel change along the axes.

## Organizational Theory and Academic Science

To this point science has been treated as an institution, which is to say, as a subsystem of society. Ideas about culture, formulated in an analysis of an isolated mountain community, have been transported without modification into a discussion of academic science. And for its part, that discussion assumes that science operates more or less independently of society. For example, the Mertonian perspective stresses the importance of scientific freedom and warns of the costs of strong societal intervention in (or direction of) science.[3] Yet academic "science" is not "society," nor does it operate independently of other parts of society. While there is a high degree of interdependence and interpenetration between science and academe, it is a mistake to assume that the goals, values, and norms of the two are equivalent. The marriage of research and teaching within the research university, for example, is a recent arrangement that limits and perhaps compromises the purposes of both endeavors. Further, academic science is especially dependent upon resources from other institutions—chiefly state and federal agencies, but for some universities the private sector is also very important. Thus the culture of academic science is a blend of the cultures of science and academe, and the resulting cultural mix is further shaped through interaction with and accommodation to its clients, competitors, and patrons. Finally, and most importantly, academic science is undertaken within organizations, and organizations have unique properties, dynamics, and goals which endure across sectors, communities, cultures, and societies.

Recognizing the important role of formal organizations in modern academic science allows ideas from organizational theory to be brought to bear on the problem of cultural change. Two potentially complementary perspectives are most useful here: resource dependence theory (Pfeffer and Salancik 1978; Thompson 1967) and institutional theory (DiMaggio and Powell 1983; Meyer and Rowan 1977; Zucker 1977, 1988). Both perspectives analyze organizations as open systems within a population or field of interdependent organizations, characterized by varying degrees of cooperation or competition with one another. The resource dependence perspective draws our attention to more or less tangible goods which are exchanged through transactions between a focal organization and the organizations in its environment. The institutional perspective, in contrast, emphasizes the organization's "institutional environment," which is comprised of "understandings and expectations of appropriate organizational form and behavior that are shared by members of society. . . . Organizations experience pressure to adapt their structure and behavior to be consistent with the institutional environment in order to insure their legitimacy and, hence, their chances of survival" (Tolbert 1985, 1). These may be complementary, not contradictory, explanations, but the synthesis will not come easily (Carroll and Hannan 1989a,b; Tolbert 1985; Zucker 1989).

Both perspectives are generally silent about cultural change, focusing instead on the structural and behavioral consequences of ecological forces. But culture and structure are interwoven: cultural innovations may follow from changes in structure and behavior, or they may occur as independent adaptations which prove just as adaptive as structural innovations. Thus, insights from organizational theory may help us understand the forces operating on the culture and structure of academic science.

Isomorphism is a force within a population that compels other members of the same environment to come to resemble one another. In their paper, DiMaggio and Powell distinguish two sets of forces toward conformity (or homogeneity) of organizational structure: institutional isomorphism, which results from competition for legitimacy and political power, and competitive isomorphism, which is driven by competition for resources and customers (DiMaggio and Powell 1983). Three types of isomorphic processes are posited: coercive, which occurs when a more powerful entity, such as the state, mandates structural change directly or through changes in rules; mimetic, which results when organizations pattern themselves on successful competitors; and normative, a consequence of increasing professionalization among organizational employees (the professionals reshape organizations in ways that satisfy the operating principles of their professions).

These isomorphic forces are readily identified in academic science (Levinson 1989). Federal regulations are a coercive force, which mandate affirmative action officers or human subject and animal protection committees, and which contribute to the creation of grants management and government relations offices on campuses. Such offices and their employees, in turn, promulgate rules and standards that change the way academic science operates. They also change the balance of power on campuses.

Mimesis plays a part, too, as universities are constantly surveying the field, modelling themselves on the successful program of their competitors by adopting innovative educational programs and forms of organization. Some of these changes may have been initiated by philanthropic foundations, commission reports, and the like. The rise, fall, reincarnation, and transformation of core curricula for undergraduate education on many campuses has its roots in such imitation. Research centers in general, and university-industry research centers in particular, are also good examples of mimesis.

The professionalization of administrators of higher education and the increasing representation of existing professions (such as law, accounting, and even such arrivistes as organizational planning and development) bring to campuses a force for normative isomorphic change as their professional standards of correct procedure replace the amateur criteria of the traditional academic administrator. This is a competition between professions similar to that now taking place in other types of organizations. In social service agencies, for example, the influx of computer professionals and experts in planning and systems management has made the social service professionals (who are typically MSWs) appear altruistic, quaint, and outmoded.

Two of DiMaggio and Powell's (1983) hypotheses are especially useful in thinking about academic science: (1) the greater the dependence of an organization on another organization the more similar it will become to that organization in structure, climate, and behavioral focus;(154) and (2) the greater the extent to which the organizations in a field transact with agencies of the state, the greater the extent of isomorphism in the field as a whole. (155) For our specific concern with changes in university cultures, these hypotheses imply that (1) increased dependence on resources from the private sector will cause universities to resemble the private sector (and, through that, to resemble one another to a greater extent as well); and (2) increased resource dependence and other transactions with government agencies will cause universities to adopt and enforce the rules and formal rationality of government bureaucracies.

In these ways isomorphic forces shape culture as well as structure. As academe adopts principles and patterns of behavior from the private sector and government, not only its structure but also its basic values and norms—key elements of its culture—are changed. Some obvious examples include increased accountability for time and resources, formal strategic planning, professional lobbying (which gives access to the congressional pork barrel), and disproportionate growth of the administrative component, with the attendant adoption of managerial standards and principles. Such changes, in turn, change the educational environment in ways that will be discussed below.

## Summary

The ethos of academic science is here conceived as a set of axes of variation. A particular university, and some "average" representing academe as a whole, may of course be located on these axes at a given point in time. But the real value of this conceptualization lies in its implication that contrasting values coexist, with one of the pair dominant and the other subordinate or recessive. Change occurs along these pathways, driven by external forces, both material and cultural, and by internal accommodations. Moreover, the ethos of academic science is a mixture of the ethos of science and the ethos of academe. It is easy to treat these as identical but important to recognize that they are not: science and academe have distinct goals, traditions, and values that have been drawn together for a time but may not remain so tightly connected in the future. In brief, academic science is situated with an organization, the research university, and is therefore subject to the general principles that shape behavior in organizations and the specific forces brought to bear by the university's environment of government, private-sector companies, and other universities. Following this logic, studies of values are led to four broad questions: (1) What are the main cultural axes of academic science? (2) How is the ethos of academic science changing? (3) Why is it changing? (4) What are the consequences of change for scientists' careers and the structure and conduct of their research?

## Overview of the Argument

The balance of this essay will draw on material from a study of academic researchers in the life sciences, described in the following section, to illustrate and examine the preceding ideas. The basic view of culture as axes of variation and the role of organizational factors, particularly competitive and institutional isomorphism, is used as sensitizing concepts, not a testable theory. In practical terms, this means that the argument is somewhat informal: it does not pretend to prove that this and no other perspective on academic science is true and useful but instead will argue that these ideas are plausible and worthy of exploration and development. Here are the basic stages in the argument.

Universities have become more dependent on external agencies for material and cultural resources such as research funds and legitimacy. These are sought from government and industry through research relations which commit the university to contribute to national and corporate goals (such as defense, economic competitiveness, and the like). The dependence is conveyed through several mechanisms to faculty and students, and it shapes their work and careers. The most prominent mechanisms of transmission are reflected in changes in the social organization of academic science, marked by new structures (administrative offices and centers), new roles (the academic marginal, the entrepreneur, the professor–employer), and new processes (changing relations within research teams and altered standards of scientific practice). Thus, changes in the university's connections with its environment have had consequences for its internal structure and functioning. Less apparent are the consequences of such changes for the culture of academic science. The "received" values of academic scientists—those values acquired during their education and professional socialization—are in conflict with the values embodied in and required by their new conditions of work. Values most strongly affected have to do with criteria for choosing research problems, appropriate working relationships with students, and standards for determining when a piece of scientific research is complete and publishable. These value conflicts create ambivalence, alienation, and anomie, which, in turn, may lead to social disorganization (including deviant behaviors such as scientific misconduct) and social change.

Having made and illustrated this argument with material drawn from interviews with scientists, we will return to a more abstract level to discuss the cultural axes (or value dimensions) of academic science. Other analysts may identify more or fewer dimensions, labeled similarly or differently; my purpose here is to initiate discussion by proposing a set of concrete value tensions that reflect three more general dimensions.

## THE STUDY

The interview material presented in this paper is based on taped face-to-face interviews with scientists in their offices. The interview schedule was loosely structured, asking specific questions while also allowing time for elaboration and exploration of the unique details of a scientist's research interests, employment circumstances, and career. Interviews lasted from one to three hours, averaging about ninety minutes.

Respondents were chosen according to several criteria. One population of potential respondents was identified through lists of grant recipients provided by NIH or retrieved from the Federal Research in Progress data base. These lists were then restricted geographically to the east and west coasts and by quality of the department where the scientist worked when the award was made. Quality was determined by the rating of a school's doctoral program in cellular and molecular biology according to the National Research Council report (1982), "An Assessment of Research Doctorate Programs in the United States: Biological Sciences." I relied on two measures: the scholarly quality of faculty, based on a reputational survey, and the influence of recent publications, determined by counting the publications and evaluating the quality of the journals in which they appeared. Middle-rank schools were within one standard deviation of the mean on these measures; top-rank schools were more than one standard deviation above the mean. Geographic restrictions were imposed to keep travel costs and inconvenience reasonable while insuring some diversity. Imposing these quality criteria on the sample allowed the study to focus on active researchers while extending its scope beyond the most prestigious departments.

A second, smaller population of respondents was obtained by asking NSF program managers to send letters requesting an interview to scientists whose proposals were declined funding in one of the three most recent panel meetings.

Respondents also were chosen to provide diversity in academic rank (ranging from full professors through recent tenures to marginal jobs), years of support (ranging from twenty-five continuous years to first-time investigators), and sex (to avoid under-representing women, especially as their positions and problems may differ from men's).

The paper draws on interviews with twenty-six scientists, two department chairs, eight academic administrators, and four agency officials. Of the scientists, sixteen were in developmental neuroscience, and ten in lymphokines (a research topic within immunology). Five were in academically marginal positions; seven were assistant professors, five were associate professors (with tenure), nine were full professors (although one of these does not have tenure and her career prospects are very similar to the marginals, despite her rank and accomplishments). They are evenly divided between medical schools and academic departments. Their educational backgrounds are generally similar in quality of the Ph.D. department—most came from very good schools—but are quite different in baccalaureate origins, ranging from the most prestigious schools to the most obscure.

Their publications and accomplishments varied substantially, as one would expect from a group so diverse in professional age and rank. But even the least accomplished had made noteworthy contributions, judged by the usual measures of publication and citation counts.

## RESOURCE DEPENDENCE AND ACADEMIC SCIENCE

> If these business principles were quite free to work out their logical consequences, untroubled by any disturbing factors of an unbusinesslike nature, the outcome should be to put the pursuit of knowledge definitively in abeyance in the university, and to substitute for that objective something for which the language hitherto lacks a designation.
>
> Veblen (1918, 124)

As Veblen's comment indicates, the current flood of work on the financial plight of universities and its consequences for educational values is certainly not without precedent. Yet that does not mean it can be ignored, despite the fact that academe has, after all, survived. To the contrary, the recurrence of themes testifies to the durability of their underlying causes and challenges us to learn from the experience.

Universities today are receiving an ambivalent message from the society at large. On the one hand, they are to preserve traditional values (and to cling onto classic texts and themes: witness the uproar when major universities change their core curricula), but, on the other hand, they are to conduct themselves in a more businesslike fashion, controlling costs and responding to national needs and the economic environment.

### Money and Academe

Academic science has responded to these expressed national needs by taking on an increased research load and, in consequence, becoming more dependent upon external research funds. For example, NIH research grants to U.S colleges and universities increased from $677 million in 1972 to $1,009 million in 1983 (in constant 1972 dollars). In the same period, indirect costs as a percent of total research costs within universities increased from 22.3 percent to 29.1 percent (Institute of Medicine 1985, 192, 195).

Among the nation's nearly two-hundred research universities, which collectively perform the lion's share of U.S. academic R&D, there is intense dependence upon the federal government for research support. Of these institutions, 70 percent perform more than half of their research and development activity for the federal government; about a third receive more than 70 percent of the R&D funds from federal sources (based on 1984–1986 data; National Science Foundation 1988, 8). Industrial support for academic R&D is much lower but has grown very rapidly, increasing from $159 million in 1972 (2.8 percent of academic R&D) to $579 million (6 percent) in 1986, in constant 1982 dollars (National Science Board 1987, 243).

Universities have changed to accommodate this increased dependence on research funds, and the changes undertaken often reflect the tension between traditional values of education and scholarship versus current demands for accountability, responsiveness, and efficiency. The most apparent accommodations are structural: the growth of research centers and institutes, the formation of faculty committees to distribute discretionary research money (a "faculty research incentive fund" or an "opportunity fund") and the creation of executive positions with responsibility for encouraging or overseeing research or "transferring" the technology produced by academic research. The titles and levels of such positions include Vice-Provost for Research, Vice-President for Government Relations, or Vice-Chancellor for Research. Entire offices may be created and staffed to serve these functions. In the 10 July 1989 issue of *The Scientist* (22), for example, one research university advertised for professional "technology transfer" positions in the university's Division of Intellectual Property, including a Technology Transfer Manager who would supervise "three technology transfer officers and seven other FTE." At many universities this would approach the staff size and salary level of a small academic department.

Another form of accommodation that reflects contrasting values appears in the allocation of discretionary funds. In such decisions universities try to honor their commitment to advance scholarship (chiefly by supporting meritorious research that is otherwise unfundable) while at the same time making shrewd business decisions (to encourage certain types of centers or to seed specific projects) and creating a reward structure that encourages scientists to pursue external research support. Meeting these inconsistent obligations often causes strains. For example, university administrators on two campuses had made large commitments to create research centers, despite opposition from faculty. Or in the allocation of discretionary grants, faculty committees must make very difficult judgments across quite dissimilar fields of knowledge about both the quality of ideas and their likelihood of attracting external support. Some projects are supported for intrinsic reasons, because they are "good scholarship," while others are supported because the idea both has merit and has distinct chances for developing into a full-scale, externally supported, research project.

Yet another university response to the changing resource environment can be found in the nature and growth of positions I call "academic marginals." Variously termed the "unfaculty" (Kerr 1963), unequal peers (Kruytbosch and Messinger 1968), and the "nonfaculty" (Teich 1982), these scientists populate an academic "never-never land" made possible by the availability of research support but made miserable by the difficulty of obtaining such support and by their ambiguous status in the university. The number of marginal scientists has grown rapidly in recent years. According to National Research Council data, academic employment for doctoral scientists in nonfaculty positions (other than post-docs) grew at an annual rate of 7.8 percent between 1973 and 1979; in contrast, faculty employment grew 4.1 percent per year during that period (National Research Council 1981, 69). Contract researchers in British universities, a comparable category, grew roughly 65 percent between 1976 and 1984, an annual growth rate of 6.5 percent (Pearson 1987; Roberts 1984).

While the origins of such positions lie in the history of the university, their growth and differentiation are fueled by the dynamics of the academic research system. In recent years the number of available faculty positions in universities has been far smaller than the number of new doctoral scientists (National Research Council 1981; Phillips and Shen 1982). For a time the surplus was absorbed into post-doctoral fellowships, and fellows were extending their stays in such positions as they waited for faculty jobs to become available (Coggeshall and Norvell 1978; National Research Council 1981). In the basic biomedical sciences, for example, the number of post-doctorals grew rapidly from 1974 through 1982, then leveled off, although the number of Ph.D.s awarded in those years grew only slightly (Institute of Medicine 1985).

These scientists are often quite capable by most measures, having received Ph.D.s from very good schools, published in major journals, and been supported by the NIH and NSF. Thus the positions are marginal to academe, but as researchers they are not marginal to the intellectual enterprise of science. Yet they are in a precarious status, suspended between graduate student and faculty member and acutely aware of it. One scientist I interviewed said, "I had a feeling when I was looking for a job that I could have fallen off the bottom . . . it's a big jump from a post-doc to getting a job. That's the crunch time. Are people staying forever as post-docs? That must be pretty weird." He is grateful to have a tenure-track job now, although he must provide 75 percent of his salary from research grants. And this is not without its stresses: "Because I'm in this . . . position I have a longer row to hoe than most—I have to be more perfect." Tellingly, he continued, "I think it's the same position a woman might find herself in," thereby calling attention to his feelings of marginality.

Some scientists in these marginal positions expressed the extreme bitterness and powerlessness classically associated with alienation. For example, in response to difficulties with career and laboratory space, one said, "I'm disgusted. I feel totally that I don't have any power and it can only hurt me to complain." Others mentioned that the system was unfair (complaining about the theft of ideas or sexist evaluations of their work); that they did not know how to compete for research grants and were not prepared for the cut-throat character of research; that the necessary mentorship was withheld from them but available to others.

Marginal positions exist and have flourished, I would argue, to ensure that academic science is more responsive to external direction and to allow universities to expand and contract in response to changes in the market. Unlike tenured faculty, marginal scientists do not have employment security to insulate them from "market forces." And unlike tenured faculty, they can be hired and fired quite readily, thus reducing the risk to the university. Marginals provide a buffer around the core activities of the university, allowing it to add personnel temporarily in areas of great current interest. In this, the hiring of marginals resembles the controversial practice of contracting out for services in the profitmaking and governmental sectors. Viewed from another perspective, under current resource limitations the traditional academic values of independence and free inquiry are in conflict with the rational (bureaucratic) values of responsiveness, efficiency, and risk reduction. One way to resolve this, at least for the short term, is to hire scientists into positions that do not embody those traditional values. But problems arise when the marginals' gratitude at being employed gives way to resentment when they find themselves working alongside "regular" faculty who have no better qualifications but enjoy the traditional academic rights and perquisites.

## The Changing Character of Academic Research

> Only by strict specialization can the scientific workers become fully conscious, for once and perhaps never again in his (or her) lifetime, that he (or she) has achieved something that will endure.
>
> Max Weber (1918, 135)

> Should a student be allowed to spend his or her entire academic career in a single laboratory learning one specialized line of inquiry or techniques at the expense of breadth in the traditional discipline? Is there a common body of knowledge, research techniques, or applications that every student should learn in a fast-moving field?
>
> Smith (1985, 9)

Federal traineeship and fellowship support have been reduced in the past fifteen years or so, their places taken by support from research projects supported by the federal government or, increasingly, by industry or state governments or by funds from the universities themselves (National Science Board 1987, 46, 78–79; Smith 1985). Students who once enjoyed the relative freedom of graduate fellowships underwritten by the federal government are now employed as research assistants with twin responsibilities to a research project and their educations.

Professors must also play two roles, teacher and employer, which may place inconsistent demands on them. (The new statuses of faculty and graduate students are reflected in such bureaucratic procedures as the changed tax status of graduate support and the regimen of time-accounting sheets, signed by the student and countersigned by the faculty supervisor.)

Not only have students shifted onto research budgets, becoming employees rather than trainees or fellows, but the character of those research budgets has changed. Perhaps the most visible change is the diminishing proportion of university research supported by the federal government and the increasing proportion that is underwritten by industry (National Science Board 1987, 78–79). There is some research showing that the restrictions placed on industrial research support influences the character of graduate education. For example, Blumenthal (1993, 1995) and his colleagues found that about one-third of their sample of biotechnology firms provide direct support through scholarships or grants for graduate students and post-docs. "Of these, one-third stipulate that students must work on problems or projects defined by the company, work for the firm during the summer, or work for the company after their training. Some universities may be concerned that such obligations place undesirable constraints on young scientists at a vulnerable and potentially very creative time in their careers" (Blumenthal 1986, 245). In a study of university-industry research relations, we found that graduate students routinely write up a piece of industrially sponsored research in two forms, one containing all the (often proprietary) details that is submitted to the sponsor, the other omitting such details (or replacing them with illustrative data) for review as the thesis and subsequent publication. Many industrial contracts are only one year long and thus do not provide the long time horizon and stability that are traditionally required by thesis students, especially those enrolled in doctoral programs. In sum, research supported by industry may pose problems that are too narrow in scope and practical in spirit to serve educational goals, may entail trade secrets that inhibit publication, discussion, and review by other scientists, or may be too short in duration to provide the foundation for thesis research.

Principal investigators on federal research grants, traditionally the freest form of support, are increasingly accountable to their sponsors for keeping their research within the confines of the proposal and for producing substantial research results within the grant period. In themselves these are not unhealthy principles for conducting research, although such departures from past practice may entail an uncomfortable period of accommodation for many. But when combined with intensified competition for research support and increased concern for efficiency (translated into tighter budgets with smaller amounts of discretionary support, shorter grant award periods, and higher expectations for the number of publications based on a project's work), these new practices are changing the working conditions for principal investigators and their teams in ways that might reduce the educational benefit of the experience.

Whatever the differences between governmental and industrial research, the important point is that conditions accompanying both forms of research support pose problems for graduate education. Decisions about the allocation of government research funds have gotten increased attention as science budgets have grown, as the importance of science for military and economic ends has become more fully appreciated, and as various crises (both real and contrived) have been conscripted to serve as justifications for supporting scientific research. In consequence, the resource environment of scientists has been deliberately changed, both through spectacular policy initiatives, such as the space program, the war on cancer, the AIDS effort, and the Strategic Defense Initiative, and through subtler shifts of priorities within agencies, such as the promotion of molecular biological approaches in the life sciences. Thus, the simple contrast between free and proactive scientists supported by federal funds, who use those funds to advance their disciplines and careers, versus directed and reactive scientists supported by industrial money (or federal contracts), who apply those funds to the pursuit of practical ends, is certainly inaccurate today and may never have been accurate.

Entrepreneurial activities on university campuses have become increasingly prominent and have attracted a commensurate amount of attention. The most visible form is the faculty-initiated enterprise, a formal arrangement that has been discussed by researchers and policy makers (Etzkowitz 1983; Louis et al. 1989). The fundamental ethical issue of such arrangements is conflict of interest between the faculty-entrepreneur's business activities and obligations as a university employee and teacher. Important as they are, such conflicts and entanglements will not be further discussed here.

Of greater concern here are the subtle and informal entrepreneurial activities of professors that come about because the university deliberately treats them as "small businesspersons" responsible for obtaining from outside sources their laboratory expenses, a portion of their salaries, and the salaries of their students and staff. In fact, the provost at one university explicitly stated that, in his view, faculty are entrepreneurs who operate within the loose confederation of the university with the administration in the role of investment bankers.

The administrative perspective on faculty is a natural extension of the university's treatment of schools and departments as "cost centers," which, in turn, may be an institutional response to the university's increased dependence on support from government and the private sector—where such forms of responsibility and accountability are routine—and to the rise of professional administrators on campus. Through such management practices universities are pushing downward the level of fiscal accountability, requiring ever-smaller units to balance their books (or show a profit). Of course, this accounting also reveals where the university's books do not balance, providing an objective, supportable rationale for budget cuts, tenure and hiring decisions, and the like. Thus, while not engaged in the splashy, high-tech entrepreneurial activity that makes the papers, many faculty must be equally entrepreneurial in their outlook and activity.

## Departmental Issues

Departments provide the daily work environments of faculty, post-docs, and students. As such, they transmit and focus the broader values and value conflicts within the university. Through departments the increased emphasis of universities on obtaining external research support is conveyed to faculty and students. And the ambivalence between educational and research goals is experienced quite clearly at the department level.

This tension between academic and research values was revealed though interviews with faculty in two departments. One department's policies forbid post-docs to work with scientists who do not have research grants, and it considered similarly restricting graduate students. Their rationale is that scientists without external research funds are not doing work that could responsibly involve a post-doc or student. This realpolitik principle has been extended to the department's educational program in the form of a proposal-writing seminar offered as part of the graduate program. (This course included a mock panel meeting with open criticism of students' proposals, mimicking the NIH peer review system.) Obtaining research support is an integral part of academic science, this department reasons, and students are well served by learning that at an early stage and orienting their efforts accordingly.

In contrast, a second department also acknowledges the importance of research funding for academic science but sees a different problem and has taken a contrary stance. Here the faculty is trying to establish a principle for allocating graduate student support that would counterbalance the influence of outside research support, reasserting the traditional disciplinary specializations and affording students more freedom of choice in their concentrations and advisors. They reason that the academic subject matter should shape students' career decisions relatively independently of its representation in scientific research, so it is essential to insulate the discipline from current research demands.

These departments have resolved the conflict between academic and research values in opposite ways, reflecting commitments to different values and conditions of academic life. The first policy prepares students for the competitive funding climate of academic science, and to the extent that "grantsmanship" contains tacit knowledge that can only be acquired through experience, this department connects its post-docs (and, through a less rigid mechanism, its graduate students as well) with mentors who have proven skills in the competition for research money. However, it risks distorting the discipline by too tightly coupling the training of future scientists with the research needs, interests, and techniques of today. On the other hand, the department that tries to insulate student decisions from the funding climate allows educational choices to be more strongly influenced by subject matter, the discipline, and individual intellectual tastes. But as high-minded as this policy seems, it risks perpetuating specialties that may be entrenched within the department but out of date within the field and makes the allocation of students vulnerable to the internal politics of the department.

## Pressures on Faculty

Perhaps the most powerful way the university can encourage grant-getting is by considering a scientist's funding record in decisions about raises, promotion, and tenure. At some universities faculty are evaluated in part by the dollar volume of research they perform in a given year. And while funding may not yet be an explicit criterion at other institutions, it is often considered an indicator of a scientist's stature in the field. For example, shortly after one assistant professor learned that his NIH project would not be renewed, about a year before the tenure decision, "the chairman made it clear that without a grant it would be difficult to recommend me staying at this institution." The message in this department is clear. "The more grants you get, the greater, speedier promotions one obtains, more space one obtains, more prestige." Conversely, "If you don't do that, you're out."

Such pressures to obtain external research support are transmitted through departments and faculty to graduate students, driven by the growing scarcity of federal and institutional support for graduate education. For example, in one department, "(When) we accept a student, we make a commitment to fund them for one year (regardless of with whom they choose to work), then . . . they either have to write a grant (or fellowship proposal) and get funded or somebody decides to fund them and they're really tied into that faculty member who doles out the money." As one might expect, this dependence introduces strains of its own.

Another faculty member whose research support temporarily ended was provided with a graduate assistant paid for by a departmental teaching assistantship. When the scientist succeeded in getting a project supported, he was told to assume the graduate student's cost, despite having no funds for that purpose in the grant. He used supply money to pay the student for one year and told her to get an external fellowship to support the balance of her graduate training. In his words:

> I kind of gave her a deadline, saying that if she's not able to find her own fund-
> ing she will have to accelerate her research tremendously in order to graduate
> very quickly because I can support her for only one year, not two. . . . Trouble
> is, graduate students, especially at her level, take a very long time to generate
> any data, and I didn't feel it quite right to pay her from the grant when she
> would presumably not generate enough results to justify subsequent renewal.

This professor was in an awkward spot, for his responsibilities to the student, in the role of teacher, were incongruent with his responsibilities to the
project, in the role of principal investigator. Stated on a higher plane, the values
associated with education and effectiveness were in tension with those associated with research and efficiency. Perhaps in an earlier time there was enough
slack in the research budget or in the performance standards of the funding
agency to allow such students to be supported on projects. But today the costs
and risks are apparently too great, so the students are compelled to seek other
sources of support or to leave graduate school.

Another professor insists that his students seek independent support for
their training: "As soon as they come into my lab I say, 'I want a grant (proposal
to NSF or NIH) for a fellowship) from you, pronto!' I tell them, 'You know, I
have to be funded. If I'm not funded, I'm out of my job. If you're not funded,
you're out of a job, so better get a grant.' . . . That's the name of the game if you
want to do anything in science now." Thus students are socialized at an early
stage into the entrepreneurial requirements of modern academic science. The
importance of obtaining research funding is also affecting the relationship
between faculty and the post-docs, graduate students, and even undergraduates
with whom they work. Tight research budgets with little discretionary money
and sharp competitive pressures (with high productivity standards for remaining supported) compel faculty to take a more instrumental view of their subor-
dinates, viewing them more as research labor than students. This may, in turn,
impede the students' educations and development of independent careers by
forcing them into the more specialized work formerly performed by technicians and by reducing the time and resources available to them for independent
investigation under the umbrella of their professors' grants. Thus the needs of
post-docs and students, who require time for independent work guided by a
professional, are not always in harmony with the needs of the senior investigators who employ them.

As internal and external support for graduate students is diminished, they
will be supported by research grants and contracts at increasingly early stages of
their education. The potential effects of this are a reduced capacity for independent work (because they become important parts of a scientist's research
effort early on) and the possibility for premature specialization as they acquire
the restricted range of knowledge and techniques required by a research proj-

ect. The pressure to specialize is intensified by intellectual change in the sciences (particularly the development of ever-more-complicated techniques), the needs and demands of principal investigators, and the tight (and relatively unrewarding) academic job market, which makes work in industry look more attractive and thus induces graduate students to acquire the most marketable skills while in school. Often this means intensive study of the technical skills (e.g., in molecular biology) at the expense of broader, substantive courses that present the classic themes and problems of the discipline.

In some cases these constraints cause scientists to act in ways that violate their own standards of proper conduct toward students. For example, a principal investigator once allowed his staff much latitude to do their own work in his lab and to participate in decisions about his projects. He found this a congenial mode of operation, bringing young scientists into the research as junior partners whose ideas were given a hearing and affording them a sense of full participation in the research enterprise. After some career difficulties, he decided to stop running his lab so participatorily. Instead, he now feels obliged to squeeze out all the effort he can on his own work, often replying to their ideas, "That's very interesting, but don't try it on my time."

Another assistant professor, who also was trained by eminent scientists, experienced a similar problem and made a similar adjustment in his conduct. When asked how much latitude he allows his staff, he replied, "Not very much. I used to do that during my first grant period. As a result, I was accused by the reviewers of my scientific endeavors of not being too focused, going in too many directions at once." So now he runs a tighter ship, which also has costs since "It delays (the grad students') educations, in essence, because I would say that the directed approach is them being more like technicians. Probably the value of their education is their ability to master techniques rather than necessarily to find out how to think, how to formulate their own problems . . . they will presumably do that at the post-doc level or as first-time faculty members."

One might contend that faculty are paid to teach and that therefore they shirk this responsibility by pressing graduate students into service in the lab. But I think this is an unfair judgment. The faculty role is now more complex, imposing inconsistent demands on the incumbents who must be businessperson and entrepreneurs, as well as teachers and scholars, thus making strategic choices that compromise inconsistent goals and values. Unlike the businessperson and the entrepreneur, however, faculty do not have profit as an ultimate standard of performance and money as a universal metric. Instead, they must respond to a broader range of goals, markets, and measures of success, absorbing the ambivalence this entails. Ironically, the economic and environmental difficulties facing the United States today may similarly reflect the inadequacies of profit as a guide to behavior.

## EFFECTS ON THE QUALITY OF SCIENCE

The competition for funding and the level of available support influence the quality of science and the character of post-baccalaureate education, sometimes in ways that are paradoxical and hard to anticipate. For example, "excellence" has long been a preoccupation of science policy pronouncements. After President Reagan proposed to increase support for basic research, his science advisor, George A. Keyworth (1983) cautioned that "the President has not allocated these growth funds to support 'next best' research" (801). Similarly, NIH Director James Wyngaarden approvingly quoted Philip Handler (1984) to the effect that "In science the best is vastly more important than the next best" (362). Despite such affirmations, 'quality' in science is a very slippery idea that is difficult to measure, particularly in prospect, from a research proposal or early in a career, for example (Leslie 1987), and is therefore exceedingly difficult to promote through policy. Indeed, I will argue that efforts to ensure that only the "best" science will be supported—efforts intended to provide high-quality science efficiently by supporting only the "best" research—may well have the paradoxical effect of reducing the quality and efficiency of science in both the present and in succeeding generations.

## The Changing Quality of Science

Changes in the resource environment are bringing about changes in the practice of science, offering students a new set of exemplars to emulate in their own work. Most prominent among such changes is the practice of partitioning a single research report into several publications, each offering only a fragment of the full account. In some instances this occurs inadvertently, facilitated by improvements in communication technology and driven by the long duration of research grants, the frequency of scientific meetings, journals' preferences for briefer manuscripts, the fragmentation of science into specialized audiences with distinct journals, and pressures from colleagues to make results promptly available. But there are also career exigencies at work, including "bean-counting" evaluations by academic administration and agency officials, who use such "objective" indicators as publication counts and citation ratios in promotion, tenure, and other decisions.

Some scientists expressed discomfort about the practice of "fractional publication" because it compromises the standards they were taught—particularly by eminent mentors who insisted that only completed and fully substantiated research results deserved publication. Their discomfort is heightened as it becomes apparent that they are communicating these new standards to their students. Now the professor has the unenviable choice of reinforcing in students a set of undesirable publication practices that are likely to be successful or a set of desirable standards that may well lead to failure; some have personally experienced the failures, such as anticipated discovery and job insecurity, caused by publishing few articles or by delaying publication to complete a fuller set of experiments. These standards of judgment and principles of research conduct are transmitted directly from teacher to student through day-to-day experiences in the laboratory and through the tangible, enduring models (exemplars) provided by the research reports and publications they produce together.

We will briefly consider two series in which the quality of science is changing. The first and most concrete sense is the quality of work performed at the lab bench, including the ability to conduct pilot studies, replicate experimental results, and assess the robustness of findings across a diverse array of substances or specimens. The second aspect of quality is reflected in scientists' willingness to take risks and their ability to absorb the consequences of risky ventures that turn out unfavorably.

## The "Bench" Quality of Science

Much of the bench quality of science—the nature of the results produced and the form in which they are reported—is influenced by the scarcity of discretionary resources and the need to appear productive in a measurable way. Scientists agree that preliminary data and replications serve to make arguments more convincing to oneself and to others. Similarly, more comprehensive manuscripts provide more room to integrate results and speculate about their implications, thus reducing some of the clutter in the literature. Yet tight budgets and intense competitive pressures remove some of the resources needed to perform preliminary studies and prepare more comprehensive papers by reducing discretionary funds and by rewarding scientists for producing more publications. These pressures were widely felt and clearly expressed by several scientists I interviewed.

According to some scientists, certain experiments cannot be done because they are too expensive, "So the result is that . . . you rely on a very narrow series of drugs—those that are cheap—that give you a good indication that you're probably right in whatever you're describing but certainly do not prove beyond a shadow of doubt that is the case." Another scientist economizes in his experiments, using two or four cases instead of six of seven. In his view, this affects the quality of the journal that will publish his results and makes him more dependent on the luck of the draw of referees to get his work published. Others have noted that the character of research publications has changed:

> Most publications have gotten shorter; the number of experiments that is contained within them is far less; the amount of information is almost miniscule;

and there's very great tendencies of people not having done quite the right thing: not used large enough samples, not worried about some of the variabili- ties that are responsible for the results that they get. So in essence almost any- thing that is reported now has to be redone by others several times before it's accepted as the truth.

This fragmentation may be caused by "the immense pressures by university authorities and by the funding administrators for proof that one is making progress. And since progress is extremely difficult to measure in terms of qual- ity, since that is a very subjective judgment . . . the other measurement is that of quantity."

## Risk

"Quality" in science means more than the straightforward quality of bench work described above; it also includes more speculative flights of imagination, intellectual leaps which, at the time taken, were hardly assured of success. Much of the history and mythology of science recounts and celebrates such flights of imagination. But changes in the culture of academic science may threaten this aspect of quality by reducing scientists' willingness and ability to undertake risky projects. This might happen if the costs of failure were to become so great that few can afford to take a chance on an idea that may be sound but does not assure success. Moreover, this aversion to risk may become institutionalized, resulting in study sections and agency officials unwilling to support a project with an appreciable chance of failure. The presentation of preliminary results, discussed above as an element of the quality of science, serves to allay scientists' concerns about the soundness and workability of their proposals. Underscoring the value of preliminary data and the risk-aversiveness of peer reviewers, many scientists say it is ideal to complete at least the first year of a research project before submitting the proposal, using the results as "preliminary" data to estab- lish the project's feasibility.

This theme arose in various guises during several interviews. One scientist regretted having been so venturesome in the proposal for an ongoing research project; the problem she posed herself was so technically difficult that good results were not forthcoming, so the publishable yield would be small and the renewal would be in jeopardy. Another warned that it is "hard enough to get stuff funded that isn't at all speculative. . . ." And still others described elaborate strategies for "attacking the problem in a somewhat more indirect way."

Scientists are not alone in their avoidance of risk. Peer review panels that recommend research support are said to avoid risk and seek mundane security. This became most apparent in the accounts of scientists who found it easy to remain supported for doing "good solid work" that was not "the breakthrough stuff." Indeed, one scientist suppressed a new discovery in a grant application, even though the work was about to be published in a prestigious journal, because he feared the study section would not believe the breakthrough and would therefore give his proposal a poor priority score.

In these examples the values of science that encourage imagination and risk-taking are in tension with the organizational and bureaucratic exigencies of the employment and research support practices of academic scientists. Practices that would satisfy scientists' commitments to their research come into conflict with those that are expedient for the university, the scientist, and his or her students. Both explicitly and by example, scientists' attitudes toward risk in research are communicated to their students. Thus, the cumulative effect on institutions, faculty, and students will be to reduce the amount of risky or speculative research undertaken and thus diminish the quality of science.

## THE CULTURAL AXES OF ACADEMIC SCIENCE

Often held up as a source of illumination on the most difficult questions and choices, the concept of 'values' is better seen as a symptom of deep-seated confusion, an inability to think and talk precisely about the most basic questions of human well-being and the future of our planet.

Winner (1986, 156)

How then can we discuss values without putting one another to sleep and without ignoring reality? We can do so by talking about how values are implemented, how they are made operational in universities and colleges and national systems. We can attempt to specify how values conflict in practice. We can search out the ways in which our institutions and systems work out accommodations among conflicting values.

Clark (1983, 14)

Chastened by Winner's warning of the potential vacuity of a discussion of values and informed by Clark's (1983a,b) intelligent suggestion of a way to inject meaning into the vacuum, we shall embark upon a more abstract consideration of values and value change in academic science. Our starting point will be the account of academic science presented and illustrated in the preceding section.

I propose that the behaviors illustrated above are motivated by several underlying ambivalences among university scientists, ambivalences that reflect change in the culture of academic science. Following Erikson's (1976) view of cultural transformation, we can try to identify these axes of change. Borrowing from organizational theory, we can look to resource relationships and the quest for legitimacy as the sources of change.

The value conflicts described above reflect tensions along several axes. These dimensions are analytically separate but interrelated, so change in one dimension will have consequences for others. Moreover, while the main conflicts are posited within a dimension, there are also noteworthy incompatibilities across dimensions. I will briefly describe each dimension and the nature of the tension it defines.

## 1. Freedom and autonomy versus accountability and dirigisme

These values define a longstanding, unresolved issue in science and science policy (Nicholson 1977). In its earliest guise the debate centered on the importance of freedom and self-determination for creative scientific work versus accountability for resources and explicit direction of scientific work, (or "dirigisme" (Rip and Nederhof 1986). Some argued that almost any infringement of freedom would impair scientists' creativity, so science must always operate independently of external influences (Stern, 1954). In the classic statement, Vannevar Bush (1945) asserted that

> Publicly and privately supported colleges and universities and the endowed research institutes . . . are uniquely qualified by tradition and by their special characteristics to carry on basic research. . . . It is chiefly in these institutions that scientists may work in an atmosphere which is relatively free from the adverse pressure of convention, prejudice, or commercial necessity. At their best they provide the scientific worker with a strong sense of solidarity and security, as well as a substantial degree of personal intellectual freedom. All of these factors are of great importance in the development of new knowledge. (19)

Others challenged this stance, even at the time, arguing that science must be closely controlled by public officials and steered into the most advantageous channels (Chubin and Hackett 1990, ch. 5). Subsequent empirical research in the sociology of science suggests that accountability, in moderation, might actually enhance, not impede, creativity under certain circumstances (Gordon and Marquis 1966; Pelz and Andrews 1976).

Today academic scientists have become more accountable to and directed by their research sponsors, even sponsors as loosely defined and academically oriented as NIH Study Sections. This accountability is reflected in scientists' choices of research problems, publication practices, the roles of post-docs and graduate students on research projects, and in the freedom and resources made available to scientists-in-training for their own work. The direction, while subtler, works directly through major funding initiatives (such as AIDS research, SDI, and major center awards) and indirectly through grant selection mechanisms that drive unsuccessful scientists into targeted areas of research. Despite increased accountability and dirigisme, and the bureaucratic mechanisms that accompany them, academic scientists still espouse and struggle to enact the values of freedom and *laissez-rechercher* for themselves and their students.

This tension also appears in the roles of academic marginals, who by the nature of their appointments are readily controlled by forces within and outside the university. Research centers and large, shared instruments also contribute to the direction of academic science by creating a level of organization that structures and regulates scientists' activities. While scientists find themselves extolling the practical and academic merits of autonomy, they also find themselves submitting to (and unwillingly inflicting on their charges) ever-greater levels of direction and accountability.

## 2. Producing research results versus educating students

Academic science has two major products: original, certified knowledge and educated, credentialed students. In the U.S. system of graduate education these two purposes have dovetailed, with research serving education (by providing hands-on experience) and education serving research (by providing motivated, skilled, low-paid labor). But the character of research has changed, driven by changes in standards of efficiency (product per dollar and per person-year of effort), expectations of utility (for the university, industrial and governmental sponsors, and society at large), and other forces, which shift the educational value of the research experience from the classical focus on free and independent inquiry to a new modality. Now the roles of faculty member (mentor) and principal investigator (employer) are becoming inconsistent, straining the incumbents. Principles and practices which the academic as mentor would prefer are inconsistent with the needs of the scientist as employer.

## 3. Local versus cosmopolitan orientation

Scientists have local allegiances to their students, departments, and universities as well as cosmopolitan commitments to their collaborators, specialties, disciplines, and the scientific community (Gouldner 1958a,b). Changing resource relationships have increased the salience of local commitments by connecting scientists' careers to activities that financially benefit the employing organization. This becomes especially salient when institutions impose standards for externally funded research activity or salary charge-out, thereby requiring members to contribute to the organization's coffers. Academic marginals are especially susceptible to such pressures as the terms of their appointments provide very little security. Indeed, such positions may have arisen and increased in number precisely because they are so responsive to changes in the environment and so expendable if they do not suit the university's future needs.

Centers and research institutes contribute to the local orientation as well, binding scientists collectively to the institution through their membership in an organized research unit which makes research awards to individual investigators from a common pool of funds. In these arrangements scientists do not obtain research support independently in their own names but collectively in the name of their institution. And suballocations to individual investigators are made by local committees based on their judgments of needs and ability. Such funds are not readily moved from one institution to another, thereby limiting scientists' mobility. Through these changes academic science may become less free as a profession, coming to resemble the profession of engineering in its growing reliance on employing organizations.

## 4. Quality versus quantity

For many scientists the quantity and rapidity of their work are becoming as important as its quality. Publication pressures and quantitative performance indicators contribute to this, but so do reductions in resources, competition for journal space, specialization, and risk aversiveness. Whatever the causes, the message is clearly understood by scientists and clearly communicated to students: the volume of work produced, not its quality alone, matter greatly to a successful career. Moreover, with resources stable or reduced (and research expenses increasing), a trade-off between quality and quantity has emerged.

This shift in emphasis serves bureaucratic needs for objective, quantitative performance standards that can be applied independently of scientists' substantive expertise. In effect, it weakens peer review as a source of professional rewards.

## 5. Specialization versus generalization

Research demands specialization, as Weber reminds us, yet education entails generalized learning. The tension may be sharpest in interdisciplinary research areas, where there is a choice between assembling an interdependent team of specialists or synthesizing disparate bodies of knowledge and their associated techniques in a single mind. The trend is toward team research with graduate students and post-docs often playing the role of technical experts, acquiring specific laboratory skills that benefit the project at the expense of their broader education. This tendency will be reinforced as training funds are increasingly drawn from research budgets.

## 6. Competition versus cooperation

Much has been written about the tension between competition and cooperation in science (Hagstrom 1965, 1974). Academic science may become more competitive as local influences on scientists become more important than cosmopolitan ones and as the competition between scientists shifts from diffuse, but certainly acrimonious, disputes over reputation and priority to concrete and career-threatening matters of resources that are easily measured and available only from known and limited sources. In effect, academic scientists, especially those in marginal positions, may become more tightly connected to their institutions and more loosely connected to the larger but more diffuse (and, in practical terms, weaker) social organizations of their disciplines, specialties, and professional societies. Such an increase in organizational chauvinism may impair interuniversity collaboration and sharpen resource conflicts by weakening one form of solidarity among scientists.

## 7. Efficiency versus effectiveness

Pious pronouncements about the paramount importance of excellence in scientific research and teaching ring hollow as pressures intensify to perform within tight budgets and schedules. As scientific work, and the work of many other professionals, is redesigned to accommodate oversight and measurement by nonexperts, it is demystified and diminished. Measurable standards of performance and detailed accountings of time and cost expenditures strip science, medicine, social work, law, and other professions of the aura of expertise that once surrounded them. Demystification is beneficial, to a degree, as a sort of populist antidote to the potentially abusive power of professionals. But it bleaches out the special character of the work, lessening its mystique and intrinsic satisfactions, finally reducing it to bureaucratically manageable units of cost and output.

## Underlying Dimensions

All of the above tensions (and probably more) can be found in the case material presented in this paper. Only slightly abstracted from the descriptions themselves, these value dimensions closely reflect the day-to-day ambivalences of academic scientists. But such a long and unstructured list is unwieldy, so it is worthwhile to seek principles that allow generalization across time and setting. Since values are nested, with transcendent principles subsuming others that are anchored to more specific times and places, it is possible to search for the more abstract ideas which underlie these concrete tensions.

In his assessment of values in higher education, for example, Clark (1983a,b) identified four value clusters from the wider society that were most salient for higher education. These were (1) competence, skill, excellence; (2) justice, fairness, equity, equality; (3) liberty, freedom, self-determination; and (4) loyalty, fealty, service to the state. Clark carefully points out that these value clusters are complex, each covering a lot of moral territory, and that they are sometimes inconsistent. For example, higher education may pursue social justice at the expense of competence; service to the state entails some loss of liberty and self-determination. He goes on to explain how higher education accommodates inconsistent values through horizontal and vertical differentiation; in effect, different standards are applied at different stages of education and at different levels of a quality or prestige hierarchy. Thus through horizontal differentiation higher education is able to accommodate inconsistent values.

The perspective taken in this paper differs from Clark's in several important respects. First and most important, the values of concern to me, while quite abstract, are nonetheless more specific to academic science than are 'competence,' 'justice,' 'liberty,' and 'service to the state.' The substance of the tensions described above can be captured in three more general value dimensions: science and scientists as instrumental versus intrinsic resources, dependence versus independence of scientists (and scientists-in-training) on each other and on their employers and sponsors, and traditional-collegial versus legal-rational authority in the coordination of scientific work. Consider each in turn.

Much of the change in the roles of scientists and their students reflects a change from treating them as intrinsic ends to expecting some utility from them. Whereas a graduate student or post-doc once might have been supported on a research grant with modest expectations about research productivity, that is certainly not the case today. And as the time between a scientific idea and its application contracts, there are increasingly high expectations that science will yield specific practical benefits rather than provide some generalized public good. Concerns about proprietary information and secrecy support this view. A prime example was the reluctance of University of Utah officials to allow outside scientists to collaborate with their cold fusion researchers until the intellectual property arrangements had been hammered out (Fuchsberg 1989).

Universities, scientists, academic marginals, and graduate students are assembled in a cascade of dependence which contrasts sharply with the ideal of independence in science and academe. While for each party this dependence serves one set of interests, it runs counter to another set of interests. For example, graduate students' short-term interests are served through service on a research project, while their long-term interests in becoming independent scholars may be disserved. Similarly, scientists (especially academic marginals) are increasingly dependent upon their employing organizations for the increasingly substantial resources to do research. And universities now rely for funds on larger organizations (federal and state governments, foundations, consortia, industrial sponsors, even lobbyists for those that partake of the congressional pork barrel) which expect specific returns on their investments (rather than simply donating to general operating funds).

The collegial and paternalistic forms of authority that are central to the brief tradition of academic science are challenged by the legal-rational authority that has been imposed on universities (through isomorphic forces) and embodied in universities' structures and rules. For example, the forms of dependence described in the preceding paragraph have given rise to new structures (such as technology transfer offices), new cadres of professionals (those who populate such offices), and new rules of conduct and standards of accomplishment which alter the traditional operating principles of academic science.

My perspective on academic values differs from Clark's in other respects as well. The institutional view of organizations, sketched above, proposes mechanisms that will cause organizations in a field under certain circumstances to become increasingly similar. If the hypothesized homogenization occurs in higher education, and I think it is occurring, then the differentiation within and across organizations that allows universities to accommodate inconsistent values will be reduced, making the value tensions more problematic.

Similarly, viewing culture as axes of variation implies that value tensions may not be accommodated or resolved but that there will always be a latent tension or ambivalence tugging on individuals and organizations, therefore always a potential for change along the axes. Rather than wonder how inconsistent values can be reconciled with one another or accommodated by an organization or actor, we should examine the transformative social forces that draw academic science toward one or another polar value, aware that such changes are not teleological but may be recurrent or even cyclic.

## BEYOND AMBIVALENCE: SOCIAL PATHOLOGY AND ACADEMIC SCIENCE

Value tensions and structural changes are neither mere scholarly curiosities nor inconsequential shifts in daily life. Instead, they can lead to profound social pathologies when individuals and organizations find themselves adrift without a moral anchor and sextant. In this section I wish to sound the tocsin about two classical social pathologies, anomie and alienation, which may afflict academic science.

### Anomie

Ambivalence poses a dilemma for persons and organizations: which of two values should be served in a particular instance? In a sense the actor's behavior is overdetermined because two inconsistent values apply to a situation, but it is also underdetermined as there are no clear rules for choosing which value to follow. Anomie, however, is a more serious state that arises when the rules guiding social conduct are weak or absent. Anomie may accompany rapid social change, as one set of rules ceases to work but has yet to be replaced with an effective alternative. Such a weakening may be underway in academic science (Hagstrom 1964).

Anomic conditions may be found in many aspects of academic science; here are but a few instances. Unfavorable funding decisions often call into question, in scientists' minds, the funding system's rationality and the fairness of its evaluation criteria. While this may be a "sour grapes" response to failure, funding decisions have grown increasingly difficult because the perceived quality difference between proposals has gotten smaller, allowing the "chance" or error component to appear relatively large (Chubin and Hackett 1990, chs. 2 & 3). Scientists in marginal positions are especially susceptible to feelings of normlessness, particularly since they notice a weak connection between their performance (which has been quite good) and the meager rewards they have received. This condition is exacerbated by feelings of relative deprivation that arise when marginals work alongside faculty whose "objective" performance is no better than their own. Thus, in the minds of academic marginals, the rules that govern the allocation of rewards are askew. Other scientists complain about the "whole lottery aspect" of the funding system, in which "Doing good work and working very hard are necessary but not sufficient conditions for getting your grant renewed." Importantly, these are NOT scientists unfamiliar with the complexities of the funding system; rather, these are people who have served on NIH Study Sections and understand their inner workings. And the young cannot rely on their elders to socialize them and to teach them the new rules, for the rules are in flux. One newly appointed assistant professor captured the failings of socialization particularly well:

> My biology professor grew up in the glory days of easy money. . . . What I don't think anybody in my generation knows is that the rules are totally different now. Those 60s rules that the old guys play and the middle-aged guys play don't apply to us. There's no easy money and there's no toleration of eccentricity. In other words, you can be brilliant but a geek and still get money in the 60s—by that I mean you can have great ideas but write a bad grant—you'd still be given money. When I was a grad student the older generation would sit around and talk about which schools in northern California they would consider working at. That mentality gets transmitted down to the succeeding generations, but those rules don't apply. I've known people who've gotten screwed, including myself, because we've had those expectations subliminally, even though rationally we know they can't be true. . . . Now you have to be perfect, you have to write perfect grants—it has to be well-written as well as brilliant.

## Academic Capitalism and Alienation

Consider these two commentaries about important changes in the structure of academic science, separated in time by some sixty-five years.

The large institutes of medicine or natural science are "state capitalist" enterprises, which cannot be managed without very considerable funds. Here we encounter the same condition that is found wherever capitalist enterprise comes into operation: the "separation of the worker from his (or her) means of production." The worker, that is, the assistant, is dependent upon the implements that the state puts at his (or her) disposal; hence he (or she) is just as dependent upon the head of the institute as is the employee in a factory dependent upon management. . . . Thus, the assistant's position is often as precarious as is that of any "quasi-proletarian" existence. . . . This development corresponds entirely with what happened to the artisan of the past and it is now fully underway . . . . An extraordinarily wide gulf, externally and internally, exists between the chief of these large, capitalist university enterprises and the full professor of the old style. . . . Inwardly, as well as externally, the old university constitution has become fictitious. (Weber 1918, 131)

In some respects, research groups in universities have become "quasi-firms," continuously operating entities with corresponding administrative arrangements and directors of serious investigations responsible for obtaining the financial resources needed for the survival of the research group. The specialization of labor in scientific research, the increasing use of highly specialized and complicated equipment, the pressure to produce results quickly to insure recognition and continued financial provision have changed certain aspects of scientific activity. (Etzkowitz 1983, 199)

What these commentaries share, of course, is the view that academic science is becoming structured along the lines of the classic capitalist workplace. Weber saw "state capitalism" in the "large institutes of natural science or medicine" (131), while Etzkowitz suggests that entrepreneurial capitalism aptly describes certain research teams. Their views are complementary, and they agree that these work arrangements have consequences for scientific research and its products.

My research supports this view of academic capitalism and suggests that alienation, the classic symptom of that form of work organization, has indeed resulted in some instances. The relationship between principal investigators and their staffs has taken on some of the characteristics of the relationship between capitalists and workers. Especially if the principal investigator is a full-fledged member of an academic department, he or she controls (but, like a corporate executive, does not own) the productive resources of science: laboratory space and equipment, research and travel budgets, legitimacy and credibility with other scientists and funding agencies, and employment opportunities. Since research staff are dependent on these resources both to make their livings and to advance from the apprenticeship of graduate or postdoctoral education into the craft of academic science, they must trade their labor for access to them. There is even an academic analog to the notion of surplus value and its expropriation: principal investigators may "expropriate scientific credit" by attaching their names to publications produced in their labs on which they have done very little work. (The language scientists use is telling on this point, with references to laboratories as "shops," and especially large laboratories as "army labs," implying a level of organization and discipline unlike conventional models of post-baccalaureate education.) These arrangements are alienating, for they separate the young researchers from their products and, through animosities and differences in life chances, from their professor/employers as well.

Anomie, the condition of weak or nonexistent norms, is a cultural barrier to the connection between values and behavior. Alienation, the condition of separation from one's products, one's colleagues, and one's self, is a structural barrier to the enactment of cherished values. Their presence in academe, combined with the other evidence of value changes discussed in this paper, suggests that aspects of academic science may be in the process of delegitimation or deinstitutionalization, while other aspects may have gained legitimacy through service to powerful interests and widely-accepted goals in society—in effect, undergoing re-institutionalization as the grounds for legitimating their activities have changed.

## CONCLUSION

Thoughtful observers of academe (Albert 1985; Clark 1983b; Etzkowitz 1983; Geiger 1986; Geiger in press; Kerr 1982; Leslie 1987; Remington 1988; Shils 1983; Slaughter 1988; Weiner 1986, among others) have detected and analyzed diverse changes in the circumstances of academe. This paper contributes to the discussion by directing attention to academic culture as a set of axes of variation and by proposing mechanisms from organizational theory, growing chiefly from the resource dependence and institutional perspectives, to explain how such changes might occur. Possible value tensions in academic science were illustrated, drawing on interview material from an exploratory study of the research practices of life scientists.

Placing the study of academic values within the larger frameworks of cultural change and organizational theory has several advantages. First, it becomes possible to generate predictions based on general principles of change. Thus we may talk with some assurance of the direction of change as well as of its specific characteristics. We can anticipate, for example, that the mutually reinforcing changes in the structure and culture of academic science now underway will continue, bringing into the university more of the practices and values of industry and government bureaucracy. The future for universities not yet on this bandwagon is less clear. They may create a niche for themselves by espousing contrary values, presenting themselves as a (non-threatening) alternative.

Second, the various complaints and warnings about threats to academic life will seem somewhat less arbitrary when assembled within a larger logic.

Skeptics can all too readily dismiss ad hoc warnings, while those that are theoretically grounded have more staying power.

Third, by thinking about culture as a set of continua defined by polar values we avoid the trap of positing an idyllic past characterized by strict adherence to a set of "good" values, followed by a terrifying present overwhelmed by a set of "bad" values. Instead, both sorts of values have always existed as potentials within a social entity; changes in conditions determine which values are expressed at any given time.

Fourth, institutional isomorphism as an explanation of social and cultural change in organizations has a noteworthy policy implication. In DiMaggio and Powell's (1983) words, such explanations should "prevent policymakers from confusing the disappearance of an organizational form with its substantive failure" (157–158). In other words, it is an antidote to the simplistic social Darwinist accusation: if some other way of running a university was so successful, why hasn't it endured?" The answer, of course, is that the transformation was not necessarily brought about by the classical competition of natural selection (that is, by failure of "unfit" organizations) but was instead driven by resource dependence and the quest for legitimacy (through such specific mechanisms as normative, mimetic, and coercive isomorphism, for example). And the forces driving change need not be as abstract and impartial as nature but themselves represent powerful interests in society.

Fifth, understanding the roots of the problem in resource relations and the quest for legitimation also suggests a remedial strategy: change the pattern of university support and intervene, directly and consciously, to modify the "contract" between the university and its patrons.

This paper is only suggestive and illustrative. To pursue the ideas proposed here would require a large scale, prospective study of the education, careers, and research activities of academic scientists. Such a study should propose a testable model of cultural change that connects values (and, more generally, culture) to the social organization and material conditions of academic science. Doing so would anchor the culture of academic science to its organizational context rather than allowing it to float freely in the intellectual atmosphere, and would invoke potentially powerful explanations of change drawn from organizational theory. Finally such a study must also consider the content of scientific research and education—not merely its social form—because the most enduring effects of changes in academic science may be found in the students, degree programs, and publications it produces.

This essay began with Weber's (1918) question and in closing can do no better than to endorse his answer: "If the young scholar asks for my advice with regard to habilitation (an advanced stage in German postgraduate education), the responsibility of encouraging him (or her) can hardly be borne" (134).

## Afterword

In a retrospective afterword it is tempting to underscore arguments that the original paper got right, excuse or ignore those that were wrong, and emphasize the currency of the original ideas. I will indulge only the last.

Since this paper was first published, much was written about changes in the social organization and culture of academic science. Gibbons and his colleagues (1994) delineated a new mode of academic science, which they characterize as transdiciplinary research conducted in a context of application in a heterogeneous style, organized in a heterarchical, transient fashion that is more reflexive and socially accountable than traditional research. Slaughter and Leslie (1997) noted the emergence of "academic capitalism" as global market forces alter the structure of academic work. Etzkowitz and Leydesdorff (1997) unraveled the mysteries of the triple helix that entwines universities, academe and the government. Clark (1993, 1995) and his colleagues studied the social organization of graduate education in comparative and historical perspective. Together these studies have documented and analyzed cultural and structural changes in the university.

The work lives and careers of academic scientists, particularly young academic scientists, have also attracted considerable attention. Czujko and colleagues (1991) surveyed young physics faculty, finding high levels of dissatisfaction with availability of research funding. In 1990 only 11 percent of respondents agreed that research funding was adequate, while 69 percent disagreed and 37 percent disagreed strongly (Czujko et al. 1991, 41). A 1977 survey of the same population reported 63 percent agreeing that support was adequate, while 23 percent disagreed, 11 percent strongly (Czujko et al. 1991, 41). This rise in dissatisfaction cannot be explained by a decline in research support for physics. In the decade from 1979 to 1989, academic R&D expenditures for the physical sciences, measured in constant dollars, rose by 68 percent (National Science Board, 1991: Table 5.6, p. 356), while over the same period the number doctoral physical scientists employed in educational institutions rose only 11 percent (National Science Board, 1991: Table 3-15, p. 294). Thus the amount of research funding available per academic physical scientist rose by about 50 percent (in constant dollars) over the period. Rising costs of research and other technical factors might absorb some of the money, but the increased pressure and resulting higher level of dissatisfaction probably result from the increased urgency or necessity for obtaining external research funds.

Goodstein has discerned a different sort of misery facing physicists and other young scientists, a misery occasioned by what he calls "the big crunch" (Goodstein, 1995). The crunch occurs when exponential growth ends, as it must when resources are limited. In the case of science, centuries of exponential growth are ending, so the newly minted doctoral scientists of a successful research professor cannot have the career of their mentor.

In the policy arena, a committee of the National Research Council (1998) has reported on Trends in the Early Careers of Life Scientists. In the main they find the 1990s marked by rising numbers of new Ph.D.s awarded, with much of the growth attributable to increased numbers of women and citizens of other countries (National Research Council, 1998, 21–23). Doctoral recipients in the life sciences are now about thirty-two years old, having spent about eight years to earn the degree (National Research Council, 1998, 27). Most go on to post-doctoral appointments, and many remain in such appointments for several years. For example, in 1995 about one-third of life scientists in their fifth or sixth year since receipt of the Ph.D. were in postdoctoral or nonfaculty academic appointments—academic marginals, in the terms of this paper (National Research Council, 1998, 48). That fraction doubled in the decade since 1985, and the number increased much more (because the number of Ph.D.s awarded rose by about 40 percent during the decade). Women of that doctoral cohort were more likely than men to occupy such marginal positions and, notably, more than a third of those who received Ph.D.s from top-25 universities held marginal posts five or six years later (National Research Council, 1998, 46).

The National Research Council committee does not, and would not, use Weber's (1918) evocative imagery, wherein "large institutes of medicine or natural science are 'state capitalist' enterprises" in which "the assistant's position is often as precarious as any quasi-proletarian existence" (131). Nor would the committee use Veblen's (1918) sputtering prose, damning "businesslike practices" in universities that would, if not checked, "put the pursuit of knowledge definitively in abeyance" (124). Instead, the committee asks in measured terms, "Should the recent changes in the career paths of life scientists be a cause of concern? Is the dismay that is being voiced by the current generation of trainees a symptom that the system is no longer optimal, or is it the normal discomfort of students reacting to the prospect of healthy competition?" (National Research Council, 1998, 64). Before turning to the committee's reply to its rhetorical question, note the phrase "no longer optimal," which implies a system that once was optimal but begs the questions of optimal for whom and according to what criteria? Times may have been better and they may have been worse, but they were probably never optimal.

In the postmodern fashion, the committee's approach to its rhetorical question is "to identify groups of 'stakeholders' who look at the current professional system from different points of view" (National Research Council, 1998, 65–68). Their responses are probing and instructive. For administrators and established researchers, they suggest, the current system works quite well. Grad students and post-docs, including those who become academic marginals, are talented, motivated and inexpensive labor. Funding agencies also benefit from current conditions, for many of the same reasons. But senior graduate students and, especially, postdoctoral fellows confront a "crisis of expectations" as they attempt the transition to full-fledged academic scientist. Evidence for this "crisis of expectations" is scant and informal, but its contours are clear and familiar. They include the perception of a large gap between scientific "haves" and "have-nots," a "pervasive sense that in the current climate of increased competition something precious has been lost," and "a widespread sense of failed expectations" (National Research Council, 1998, 68). While the NAS does not use the terms, such conditions may also be termed *alienation* and *anomie*, the former referring to the separation of a person from self, others, and work, and the latter to a condition of weak or inconsistent rules guiding behavior.

Much has changed since Weber and Veblen wrote about academic life. The scale and scope of the academic research enterprise has increased tremendously, and the drive for science to engage practical matters has probably intensified, through the increased involvement of the private sector and through direct government action (e.g., the Government Performance and Results Act of 1992). The composition of the academic workforce has changed utterly, including women and persons of diverse ethnicities in proportions unimaginable in the early decades of the twentieth century. Academic science has professionalized greatly, with more regulations (e.g., human subjects and animal subjects protections, hazardous materials regulations, financial accountability rules) and more professional societies. Scientific communication has changed, too, with explosive growth in the journal population about to be eclipsed by more rapid and explosive growth in electronic means of publication and communication. Disciplines and specialties have been created, have differentiated, and have been recombined into new interdisciplinary specialties, such as structural biology, which uses computers and sophisticated software to apply physical understandings about the structure of molecules to explain biological phenomena.

Yet for all these changes and more, the warnings issued by Weber and Veblen remain eerily true today. Academic life remains precarious, suggesting perhaps that for all the change that has taken place, remarkably little imagination has been applied to rethinking the fundamental organization of the university. For all the changes in the substance, organization, and politics of science, remarkably little has changed in the personal experience of the scientific career. Science still demands a vocation, and those without a calling enter at their peril.

## NOTES

This paper was prepared for the Project on Values Training and Ethical Issues in the Graduate Education of Scientists and Engineers, conducted by the Acadia Institute with support from NSF Grant No. BBS-87-11082. The data reported here were gathered with support from NSF Grant No. PRA-85-14061 and a grant from the Paul Beer Trust at RPI. An earlier version was drafted for the Office of Technology Assessment under contract H3-4075.1; recent writing has been supported by NSF Grant NO. BBS-87-11341.

This and earlier versions of the manuscript benefitted from the helpful comments and criticisms of Tom Carroll, Daryl Chubin, Mark Frankel, Sharon Harlan, Deborah Johnson, Sal Restivo, Robert Silverman, Langdon Winner, Ned Woodhouse, and two anonymous JHE referees. Therese Landry carefully edited the penultimate draft. I am especially indebted to the scientists who participated in the study, freely and cheerfully discussing their work and careers with an outsider whose purposes must have seemed obscure.

1. This is not intended to be naive, dichotomous thinking; the value poles represent ideals, not practical alternatives. A given institution may be located at any place along the continuum, although logical and emotional appeals may be expressed in terms of polar values.

2. This argument parallels Erikson's work on Appalachia (Erikson 1976, 51–78), in which he described how the received view of the "mountain ethos," derived from the brief visits of prominent commentators to the mountain communities, is for predictable reasons a one-sided and incomplete account.

3. These emphases also reflect the times of Merton's writing: the events in Nazi Germany were at the forefront of everyone's mind. This appears most clearly in "Science and the Social Order" (Merton 1973, 254–266).

# 6

# Building Labs and Building Lives

JENNIFER L. CROISSANT AND SAL RESTIVO

The broad aims of the study we report here were to explore the values and
ethics issues which arise in the context of industrial and government sponsored
or collaborative research and to document and discuss organizational changes
in the university environment driving and affected by these interactions. Our
research, one particular form of UIRR interaction, was part of a larger study of
research centers at Rensselaer Polytechnic Institute (RPI). The first sections of
this report deal with three laboratory sites: the Sailplane Project, the High
Temperature Advanced Structural Composites Lab (HiTASC), and the Chemical Vapor Deposition (CVD) Lab. All of these are part of a larger Center for
Composite Materials and Structures (CCMS), funded in large part by the
Defense Advanced Research Projects Agency (DARPA) and the Office of
Naval Research (ONR). The penultimate section concerns the Metrology laboratory, where we focus on industry and academic interactions in graduate education. Across all these accounts, we are centrally concerned with the idea of
'manufacturing,' that is, the production of people, places, things, and institutions.[1]

In the context of exploring government (meaning here predominantly military), university and industry interactions in the United States, the HiTASC
project provided an opportunity to study the interactions of a military agency
with an academic institution. Even this interaction, however, is framed around
attracting the attention and cooperation of businesses. In the Sailplane project,
the theme is centered more directly on industrial activity and undergraduates,
although there is a significant history of military influence on the emergence of
the project. Our stories examine discourse and conflicting values and the social
relations of participants in and among the engineering laboratories. Among the

characters in the following sociological stories are professors, students, staff, institutions and agencies.

We should clarify the basic materials of inquiry for the laboratories in question. Composites are usually manufactured from polymers and epoxies or various plastics and strategic fillers, such as carbon fibers or filaments in a matrix, fabric, or strand. Often composites are laminates, or layered arrangements of synthetic materials. Composites are generally lighter in weight, or, more specifically, have higher strength-to-weight ratios, than conventional metal alloys. They are new materials; few components are found in nature. Composites are highly desired by both producers and consumers of various goods. However, complete characterization of the materials, their safety, wear, fatigue, behaviors under stress conditions, and their complexity of manufacture have slowed the adoption of these materials. These complexities, and the desirability of composites because of their material properties, have made them interesting to industries and the military. They are desired primarily for aerospace applications but are used for other systems as well.

The metaphors of shaping and molding focus our attention on general processes for socializing students into academic, industrial, or government values. The DARPA sponsorship of the HiTASC facility is the material manifestation of a social world of graduate students learning to orient their work toward military uses. We had the opportunity to observe actual laboratory construction. The Sailplane project has military and industrial forms of order, particularly hierarchical; this is the context for pursuing commercial interests in this undergraduate academic environment. Here we observed the construction of students as well as an airplane. Students are shaped and molded through their interactions with the organizational and material worlds around them, and to some degree, as participants, they have the chance to shape their environment and its meanings. In the discussion of the HiTASC and CVD laboratories, we consider the construction of the material environment, the laboratory space, and its expected influence on students and the intellectual work of the lab. The building of a laboratory is an exercise in institutional politics, a negotiation of tangible and intangible boundaries and priorities. The opportunity to build a learning environment for students provided an opportunity to reinforce or demand certain styles of working and interacting. For the Sailplane project, the goal was building students. In both cases, design decisions are indicators of what and who counts, of core values and constituencies. As we relate below, conflicts and crises also provide indications of operating values.

## PROJECTS, PLACES, AND ORGANIZATIONS

In this section we point out some of the important characteristics of the settings, which emerge from their institutional histories, and briefly discuss our

research methods. The HiTASC project is organized in two adjacent and connected laboratory spaces with overlapping sets of actors associated with each
lab. The Sailplane Project is centered in a single laboratory space, although students often do computer-related work at other sites on campus. To recall our
earlier description, both projects are subdivisions of CCMS, an interdisciplinary
organization with researchers and students from several materials and engineering fields.

## DARPA/HiTASC

Our story of the DARPA/HiTASC program begins with an excerpt from an
August 1987 annual report. The research contract was awarded in 1986. This
leaves unexamined the processes that predate the award of the grant and only
summarizes the first year, but it is a convenient place to begin. We were not
given access to original contracts between RPI and DARPA. The research
objectives of HiTASC are found in the report and summary documents. These
objectives are linked to specific goals:

1) An educational thrust to develop an integrated course sequence to provide . . . graduate research assistants with an appropriate, comprehensive,
   and fundamental education in the field of high-temperature composites.
2) Research goals of: a) Identifying potential high temperature composite
   systems; b) developing and understanding new materials and processing
   routes for fibers, matrices, and their composites in both ceramic and
   intermetallic systems; c) understanding the thermomechanical behavior
   of high temperature composites at both the micro- and macro-mechani-
   cal levels from combined theoretical and experimental approaches.
3) Establishing a unique national resource for the benefit of the United
   States.

Five research program thrust areas are identified around these objectives, which
generally appear as excerpts from the three goals stated above, with the addition of the "installation of a state-of-the-art fabrication and test facility for high
temperature composites." These facilities are the Chemical Vapor Deposition
(CVD) and Advanced Composites Laboratory (ACL). Before getting into the
specifics of the HiTASC projects, and the laboratory construction, the context
for the Sailplane project should be established.

## The Sailplane Project

In stories of manufacturing in higher education, students are the ultimate products. *Students*, as a category, product, and commodity of and in changing university and industrial collaborations, has a variety of meanings. Students are

certainly not passive participants in the educational process, but often are interpreting their education in ways unsuspected by the faculty or industrial sponsors. From an economic perspective, students are a means of production and a technoscientific resource for industry. The Sailplane project is a way of exploring these issues. The Sailplane lab is located in the Composite Materials Laboratory, a basement lab in the Jonsson Engineering Center (JEC). The Sailplane Project is an enterprise "involving design, fabrication, and testing of an advanced, full-scale, flight-worthy all-composite aircraft" and is "a highly productive way to conduct research on processing science and technology for low-cost aerospace structures" (interview archives). Administered under the organizational umbrella of the CCMS, the Sailplane is referred to as RP-3 in most promotional publications. Sponsored by as many as ten industrial firms and three government/military agencies (NASA, AFOSR, and the ARO), the project is also referred to as CAPGLIDE (Composite Aircraft Program Glider) in NASA-oriented texts and internal reports. Military funding for composites research began during the 1960s. Initial projects in composites work were begun with a NASA grant in 1960 as well. The university's initial activities were in aircraft components (such as tail or wing sections), research, and prototype manufacturing, through the academic organization of the Materials Research Center (MRC). Civilian aircraft interests were incorporated into research agendas in the following decades. The current project is purportedly designed to attract the attention of industry as well as continue with military applications. Lighter, less costly, more reliable, and longer-distance aircraft are needed to adapt to the changing military airbase infrastructure. Improved fuel efficiency and high technology are cited as the needs of the civilian air interests, especially for dealing with foreign competition. Additionally, the university administration sought, and continues to seek, prestige research that will attract media attention. A high-visibility project such as the Sailplane has this quality.

The RP-3 is the third in a series of full-flight projects. The RP-1 was performing flight tests in the summer of 1981. The RP-2 was under construction at that time, and its initial flight took place in the following year. A design contest was initiated in the spring semester of 1986, and the winning design was to be the basis for the next iteration of the sailplane. Occasional flyers appear on campus bulletin boards, advertising for students to become involved in "The Sailplane Project," but these do not specify whether this is for the completion of Sailplane 3 or the initiation of project number four. We discuss this ambiguity below.

The fieldwork in the Sailplane lab took place between late February 1988 and April 1988. A participant/observer who had spent two semesters as an undergraduate in the Sailplane lab the year before this study was initiated carried out the work under our direction. Fieldwork in the HiTASC lab was carried out between September 1988 and early May 1989, and in the CVD between mid-October and mid-May, with an on-site observer spending approximately one day a week in each of the laboratories. The Metrology lab had a similar schedule for participant observation, although the work we present here focuses on a document trajectory, the development a laboratory contract, with some attention to the resulting research activities.

As studies of science and engineering practice, our accounts are installments in the accumulating microsociology or ethnographic study of laboratory life.[2] Unlike most, however, we are concentrating here on the educational components of laboratory activities. This chapter is also meant to add to the repertoire of available materials on scientific and engineering practice, and to fulfill the goals of previous studies, namely, to describe, demystify, and contextualize science and engineering. Finally, the work here is offered in the spirit, although not the methodology, of other work on the socialization of scientists and the active engagement of students in constituting themselves as actors in "communities of practice" (Jacoby and Gonzales 1991).

## Building Laboratories

The built environment can be seen as an embodiment of various technoscientific ideas, such as environmental control, and broad cultural norms, such as safety.[3] The devices and materials—and the negotiations over them—are central to illustrating the values and expectations of participants. The development of the DARPA/ONR HiTASC facilities was an excellent opportunity to explore how values are embedded in built form, and what values teachers and sponsors expected to be transmitted to students in that environment. We are particularly concerned here with the ordering of day-to-day practice, the norms for what is considered "good work," and the processes of building values into materials, practice, and people.

The symbolic and material barriers between public and private, patterns and rights of access, and the expected social standing of practitioners and observers of science are apparent in laboratory design (Shapin 1988). We do not espouse a naive architectural determinism. While the laboratory designers may have had certain goals and expectations, much of the interpretation and negotiation of the facility is left to the students and faculty in practice. Certain behaviors are constrained by lab walls and equipment, but much is underdetermined. We also need to consider the selection of specific instrumentation, which will greatly shape the future of research in the laboratory as well as the gross architectural components of the construction process.

As outlined above, a primary "thrust area" in the HiTASC project is the construction of a testing facility. With this facility, the development and testing phases of research and development are located firmly in the university setting. While we are not interested in the debate about defining applied and basic science, or the appropriateness of various settings for these activities, the specific construction of a testing facility is problematic. With the HiTASC facility, the initiative for the laboratory was apparently taken by the granting agency and welcomed by the university researchers. It does seem to be a shift from more conventional research identification and allocation patterns. It also challenges the educational agenda, which we discuss more fully below. Whether it is a significant challenge to traditional academic patterns and values, the retrofit of a building to accommodate a laboratory is a fascinating opportunity to study the built environment and its embodied values.

The CVD and Advanced Composites Laboratory (ACL) are in the Jonsson Engineering Center. These laboratories were moved from previous locations in another building as a result of the DARPA contract. The need for more space generally, and for the specialized testing facility specifically, provided the impetus for the move. This entailed a change in symbolic and institutional boundaries, as different research groups and departments came into contact under the umbrella of the CCMS, and a change of physical boundaries that had to be constructed or renegotiated. Composite materials was solidified as a research area with the development of the laboratory facilities. The Center was similarly stabilized as an interdisciplinary unit, as was a consensus on appropriate research topics and general patterns of personnel recruitment and activity.

There were two crises in the DARPA program during the course of this study. One involved the impending (March 1989) site visit by the sponsoring agency. The second was the theft of the computer. Later, we discuss these events in terms of what they indicate about cohesiveness and group conviviality; here we use these events, particularly the site visit, to highlight key values.

The site visit was perceived to be a crisis because the HiTASC testing facilities were far from complete due to considerable delays in setting up the laboratory. In November 1989, three years after the contract award, the laboratories were incomplete and equipment was absent or nonfunctional. The move from the original facility in the MRC to the new JEC disrupted research and patterns of work and collaboration. The HiTASC project ran chronically late. The 1987 annual report was filed approximately one year late. The March 1989 site visit led to a flurry of activity. Not much research had been carried out, but the laboratory needed to appear organized and functional.[4] In the end, the DARPA/ONR visitors to the university were given an incomplete tour.

From the electricians' perspective, delays were the result of continuing changes, expanding expectations, and poor initial design. Changes in specifications and in the physical orientation of the laboratories were often made on a daily basis. In one instance, the decision to add another furnace required a major redesign of power connections. The laboratory was designed to use an enormous amount of electrical energy, and many devices required high currents and special lines. Instrumentation changes required infrastructure changes, at least until the point where the infrastructure was stabilized and thus interpreted as a constraint on the instrumentation.

At one point an electrical designer and co-worker were making sketches on a yellow legal pad, with no formal blueprints or guidelines for the project to be found as of November 1988. The electricians left the laboratory in early 1989, citing a lack of work; they could not continue with their installations until additional instrumentation appeared at the facility, and there were insufficient funds to purchase the equipment and pay the electricians. A student in the lab cited poor planning, unrealistic goals and estimates of time and cost, and ad hoc changes in the plans. The project was complex: systems needed to be synthesized nearly simultaneously. Electrical, safety and ventilation, plumbing, information, and the supply of parts and research equipment all had to be very carefully coordinated. At the March 1989 research sponsor review meeting, leading faculty implied that the lack of laboratory readiness led to losing a contract with an industrial sponsor. The program director's secretary remarked that a staffing shortage led to the delays. A research assistant and administrative assistant were supposed to be helping with laboratory set-up and guiding students in their research. The lab director was to be responsible for both the development of the laboratory and the supervision of his graduate students and the personnel in the center more generally.

A December 30, 1988, memo from the project director to the acting dean of the School of Engineering described the costs to the program of an October 1988 computer theft. It also mentioned the losses due to the move: approximately $1.2 million of research "output" and another $100,000 dollars of additional contracts, plus losses of prestige and poor performance on a contract with DuPont and other industrial contractors. A January 23, 1989, memo to the director of physical facilities was also sent to the acting dean and other administrators requesting a concerted effort to complete the laboratory by February 24, 1989, because of the upcoming March 3, 1989, visit. The laboratory's condition for the site inspection threatened "continued funding" and the potential for future awards and research progress.

These conflicts and perspectives were partially resolved in the near-term of the site visit with a large amount of graduate student sweat. Equipment manufacture was accelerated, and a considerable amount of energy was expended in a cosmetic construction of the lab. Desktops were cleared, cabinet doors put in their place even though the ovens and devices that were to be put into the cabinets were missing or not functioning. Desks were organized, and wires, plumbing, and coverings for devices tidied up. Photos illustrate the dramatic changes in the appearance of the facilities, despite the nonfunctional equipment, precipitated by the site visit. The lab approximates what a working lab is expected to look like, with the instrumentation and floor layout well ordered and clean. This was the public face of the lab, although the working face for the development of the labs was one of nearly complete disorder. The established labs in use in the building had, of course, a much more lived-in look.

The boundaries and contents of the lab are examples of the politics of research groups. The people in the lab are a part of an organizational structure and a social group. Outsiders are suspect, as with any close group, but what is of interest here is the use of resources by nonlaboratory members. Students used machines in other labs, and students from other research areas came to use the working ovens and furnaces in the CVD and ACL. Two researchers at the DARPA/ONR 1989 site visit sessions mentioned the trials and tribulations of buying time from other departments to use facilities. On another occasion, students discussed the costs of training to use the facilities in another laboratory group in another department.

These translaboratory connections are important in the general institutional scheme, but they indicate in some sense who counts as an insider. One particular graduate student, who was at the time of the fieldwork the only one with an active experimental research project, one not requiring the still out-of-order equipment, was not in the laboratory group. His work was funded by General Dynamics; he worked in both the MRC (old space) and the JEC (new space) and could have waited for some of the new, unfinished devices but instead completed his thesis using instruments available in other laboratories. The student was independent of the others, did not spend much time in the lab, and did not need to work on the construction projects for which other students were responsible. This relatively minor case indicates potentially growing problems of internal accounting, ownership of materials, and the potential for redundancies and diseconomies of scale as laboratories duplicate equipment to preserve group boundaries.

Besides the laboratories providing physical boundaries for the group, the CVD and ACL spaces exemplify some other norms, such as access, privacy, and safety. The professors' offices are not directly connected to laboratory spaces. The laboratory is closed from public view, and there are no windows to the outside. However a window between the labs provides for communication and surveillance across the spaces. What private space is available to the students is limited to desks and behind partitions, while workspaces at the bench and the computers are in collective areas. Student safety has some value: safety systems include "sniffer" equipment for detecting toxic or flammable gases. Gas exhaust and waste management also reflect environmental concerns, although the installation of these systems was delayed and students began their research before the systems were completed. The furnaces in this lab for curing composites are water-cooled to prevent overheating, with safety switches to control the flow of water. The lab director pointed out in March 1989 that "these switches . . . will not be available from the distributor until April 27. They [the students] will have to run the furnaces without the switches at their own risk."

The CVD laboratory setting is inflexible. Partitions and furnishings are immovable, as is the orientation of various instruments. As the electrical designer for the institute remarked, the "whole place is an instrument." Changing research foci or apparatus will entail significant labor and expense. There is a built-in research inertia in devices and their configuration. The ACL is somewhat less structured, although bulky equipment does impose constraints. The selection of devices implies the selection of knowledge. As noted earlier, one student had no problem making steady progress on thesis research. He used relatively simple ovens for curing composites, while the other students' projects required complex vapor deposition or atmospheric ovens. The program director, whose specialty is chemical vapor deposition, designed the laboratory to fit his research agenda. Undoubtedly it will long outlast his tenure at the Institute.

## Building Lives

The focus of this section is the construction and negotiation of lives, the socialization and group-building processes within the laboratories. Records of day-to-day interactions, as well as materials from annual reports and presentations provide evidence of the building processes.

*DARPA, Composites, and Research Identities.* We will begin the DARPA/ONR HiTASC story with a retreat of faculty and agency representatives and the 1988 annual report. While the group is solidifying into a composites research center, traditional disciplinary boundaries are under negotiation. As stated in the report: "Weekly seminars force interdisciplinary interaction and enrichment for the students and faculty. These meetings have been successful in helping develop a 'program mentality' for the group, especially at the student level." "Students working in a multidisciplinary area have to have some understanding of all disciplines involved." And "To make immediate contributions when they graduate, students must learn interrelationships between research, technology, and systems." A great deal of attention is paid to the students as "outcomes," reinforcing the role RPI is to play as an educational institution. The president of the Institute appeared at the 1989 annual presentation to the sponsoring agency and made comments about RPI as a successful institution on various levels, particularly in its attention to industry: "Students are being taught well to make links to outside industry."

Students were very conscious of these issues, and not just about outside links to industry, but also of the problems of patronage in academic research. For example, after the 1989 site visit, students were joking about funding and relationships to sponsors. One student, who was responsible for getting a visitor to the airport, quipped that they were worried they would lose funding because his car was dirty and he wasn't a very good driver.

Students were also aware of institutional politics as well. They were found discussing the costs to the department of sending them to another department for training on some sophisticated machinery. While somewhat awed by the cost, the students also capably noted the institutional networks the project director was manipulating to get them access. At the time of the study, there were fourteen faculty, four research engineers, seven post-doctoral associates, twenty-one graduate student research assistants (M.S. and Ph.D.), three fellows, and six undergraduate assistants in the CVD and ACL groups. The director of the program had two postdoctoral associates under him, as well as five graduate students and three of the six undergraduates. The influence of various students, access to resources, and kinds of tasks followed an implied hierarchy. While the organizational chart visually displayed arrays of faculty, post-docs, and students, each equal within their level, the practical organization reflected the relative power of researchers in terms of access to students and funding, while students' statuses were related to their association with a mentor.

As the program director commented: "Our main product is students . . . engineers, students who think of composites as they would of conventional materials." In some sense, this was readily apparent. Already at the research level, specialization was heavily reinforced. A student desired research work in polymers, but was doing fiber chemistry for composites work because of an assistantship. So, not unlike other fields of academic work, students in this lab were organized in specific and important ways. Students can, of course, adapt to or adopt other forms of social organization, or change fields. However, as they initially walk out of the laboratory door with diplomas, they are most familiar with a specific form of organization, a well-defined set of intellectual and practical skills, a set of assumptions about relations with nontechnical staff, and a very strong set of experiences and beliefs about behaviors toward patrons and research sponsors.

One other important point is worth emphasizing. As noted earlier, one of the thrusts of the program was the development of an advanced materials curriculum. The explicit incorporation of an educational agenda in a research contract with DARPA is interesting. Progress toward the educational program goals is stated in the annual reports. This opens up a research question as to the extent to which contracting courses into the curriculum can be a significant factor in disciplinary expansion at institutions.

Statements about education in annual reports and elsewhere serve to reaffirm institutional boundaries.[5] This allows participants to make claims on resources, about utility, and to impose constraints. Curricular changes attached to research grants may be somewhat problematic to the spirit, if not the reality, of an intellectually independent university.[6] These program changes had obvious results in shaping the educational agenda: there were nine courses offered in 1989–1990 in composite materials and engineering. Four of the courses were new additions. It is difficult to track the specific effects of DARPA/ONR influence on enrollments and curriculum over time, although the composites program generally had expanded to nearly fifty enrolled graduate students at the time our research was completed.

In the first annual report to the granting agency, the CCMS emphasized the educational benefits of the program. There were nineteen scientific reports presented. It is unknown whether the progress on the laboratory construction facilities was discussed, although that had no formal place on the agenda and in the presentations. In the 1989 report to the agency, only eleven scientific presentations were offered, although the list of presentations, publications, and interactions with visitors was longer than that of the previous year. Much of the prefatory viewgraphs and materials were identical to those in the first year's report. The emphasis on education was expanded, and regional competition on composites research was stressed. Appended to the 1989 presentations was a new wrap-up, which discussed expenditures and summarized the research in a series of problem definition/solution statements. The report format is highly ritualized and, not unexpectedly, presents the best possible face to agency representatives.

The DARPA project has generally been viewed as successful in terms of projects completed, degrees granted, and presentations delivered. Various objectives, such as laboratory construction or specification and analysis of novel composites, are evaluated and placed in the context of the large-scale goals. The reports and their "progress talk" have specific functions for groups in terms of self-definition and justification, as institutional capital for negotiation, and for group morale. The Sailplane project illustrates similar characteristics.

*Design or Assembly Line?* The RP-3 Sailplane functions as a sacred object around which activities and identities are organized for a group of students. It serves for the larger Institute as a symbol of what young engineers can do. The first full-aircraft project, the RP-1, was completed in 1981. Two years later, the second craft was undergoing flight tests, while further modifications and tinkering for the optimization of flight capabilities continued on RP-1. Until 1980, undergraduate work in aerospace components focused on manufacturing prototype parts, such as internal framing and support struts. The Sailplane Project began as the manufacture of aircraft components specified by Boeing. This appeared to be uninteresting to students, and its educational merits were questioned. It appeared that students were doing assembly work, not design as befits engineers as opposed to technicians. The Sailplanes were justified on the basis of educational content, research and design experience, and academic validity from the perspective of the faculty researchers. Students still spend much of their time in assembly; however, now they construct parts of a plane of their own design, rather than components selected by Boeing. The data or science resulting from students' efforts are now one step removed from parts testing protocols of the earliest projects. The framing of composites manufacture as supporting student-centered design was changed to reinforce its image as an educational process rather than an assembly line for parts testing.

The design contest for RP-3 was announced in late fall of 1985, with a late spring 1986 deadline. There was one entry. In the fall of 1986 analytic design began, including computer modeling, drawing, calculation, and documentation of design decisions. Although it was contingent on testing and fabrication results, the design process was completed in the spring of 1987. Students mentioned selecting the project for several reasons. They could do "real design," have an independent schedule, learn (especially whether theory was useful), and aid their career advancement. Students could also receive credits and bypass a notoriously difficult laboratory course required for graduation.

The workspace for the sailplane was filled with posters of Boeing aircraft, the space shuttle, military aircraft, and many pictures of gliders. There were several magazines on gliding and numerous pictures of successful projects from the past. It is a busy space, cluttered with devices, spools, and containers of materials, but not disordered. There is little personalization of the space, with no private work areas for individual students. Only the project supervisor has office space.

At the time of the observations in 1988 there were two rather distinct kinds of students: about a half-dozen "old" students present since the design process of 1986–1987, and "new" students, often viewed as transients by the established ones. Old students will be the ones most likely to receive credit if and when the project is finally completed. Some new students may become old students merely by staying with the project long enough. More of the 'old students' may graduate or otherwise leave, necessitating the recruitment of new students for continuity. Old students were more comfortable and free-ranging in the laboratory space, and could even enter the supervisor's office in his absence. They clarify design decisions and help and instruct new students with the construction processes. Among the established students, two in particular demonstrate leadership in the laboratory; they spend more time in the laboratory than any other student.

All students needed to be familiar with the preparation of composite laminates. Students were to prepare a "lay-up," a small section of composite laminate consisting of a sandwich of epoxy, Kevlar fabric, and graphite fabric. Students assembled these materials in a mold, then placed the configuration into an oven under vacuum to remove air from between the layers. This lay-up process insured that all students, even those doing computer analysis, had at least some familiarity with the capabilities and limitations of the materials, so that they would not design impossible-to-fabricate or nonfunctional parts.

For the first group of about twenty-five students embarking on the initial RP-3 project, lay-up processes were accomplished in the first three weeks of the project in fall 1987. The process was reported to be very sociable, egalitarian, relaxing, and interesting but not over-taxing in terms of manual work. Students reported listening to the radio and having conversations in a friendly atmosphere. This first group of students, or what's left of them, is the core group of "old students" at the time of the fieldwork. They completed the analytic design in the following (spring 1988) semester. Students were expected to contribute ten hours per week and attend weekly meetings. These meetings at first focussed on system design, discussions of assembly techniques, FAA information, airing grievances, and progress reports from design subteams.

After the initial experiences doing lay-ups in large groups, the students subdivided into design teams. Several ended up doing computer work and were "rarely in the lab." New, structured and hierarchical interactions were incorporated: groups, managers, and meetings appeared. The preliminary design group contained nine students plus a student manager and was broken down into subteams of aerodynamics, performance evaluation, and stability/control. Other groups worked on engine specifications, landing gear, and cockpit design and controls.

For the old students the design experience was to be about consensus-based decision making, at least in theory. Of course, with individual influences and managerial politics interfering with consensus processes, there were lapses. Norms of practice, both organizational and technical, arose from the interactions. Students kept time sheets, several worked on compiling PERT/CMS scheduling documents, and so on. A sense of protocol for communications and design professionalism emerged. For example, one group optimized sailplane stability at twenty-six feet, while other groups had reasons for adopting a different value. As one student remarked: "I always thought they were forgetting we were the preliminary design group, and there is a place to start designing a plane; surely not at the landing gear." There were complaints of going over manager's heads and incomplete communications disrupting the design process.

In the idealized world of engineering design, the politics of group negotiations are seen as aberrations, deviations from "real" design, rather than inherent in, or perhaps more accurately, constituting the design process. This is a form of decontextualization not undone by students' educational experience.

No doubt, they gained skills in negotiating design decision making and group interaction and management.[7]

In the continuing years of the project, most of the work entailed constructing, treating, testing, and fabricating panels and parts, much of this purely for practice. Student logs are long lists of sanding, cutting, gluing, assembling, testing, and fabricating processes. These manual tasks took up to ninety hours per student over the course of the semester. But not all students completed a journal or the semester. Only the six or so old students continued to work on analytic design. New students also worked alone more frequently, since they were entering the project not as a group like the old students did at the beginning of the project, but as individuals. Unless new students become old ones, they are not likely to become involved in design decisions. As the project continued and design solidified, new students had fewer opportunities to reopen or revisit design decisions, and were mostly occupied with part fabrication activities. Posters and advertising for the project signal a high turnover of student personnel. Memoranda between supporting faculty and the project supervisor illustrate some concerns with the current project, in terms of "student generation" time. That is, a project must be completed in three years or less, so that a group of students can start and complete a project within a typical four-to-five-year course of study. Increasingly complex projects make it difficult to get students. Deadlines were viewed as overly optimistic, particularly the goal of completing projects within two years (spring 1988). Follow-up research indicates that there have been significant delays in the program. As noted above, flyers were repeatedly found across campus, but it was not clear whether this was for Sailplane three, or for the initiation of a fourth project. As of 1997, the RP-3 sailplane had not yet flown, although Web-based press materials indicated that they were hoping that a noted experimental aircraft pilot would fly it. The RP-3 flew a twenty-one minute test flight at Schenectady County Airport on December 9, 1999. The related press releases noted the 1988 project start, and that approximately 1000 students have had contact with the craft in design and laboratory classes. Press materials also suggested that a fourth design might be initiated in spring 2000.

Despite this delay for the sailplane, at various meetings and presentations to students and tours visiting the lab, the director of the program cited several major positive attributes of it for students and for the field of composites. The first benefit is the gaining of understanding of the strength and stiffness characteristics versus weight properties of various composites. For aeronautical activities, weight translates into fuel use, so maintenance or improvement of material strength characteristics with weight reduction entails fuel and cost efficiency gains. Second, information, specifically manuals on composites assembly for the military and civil air industry, is part of the program benefits. Documented in reports and manuals, experience becomes codified as efficient and reliable design and manufacturing techniques. Additionally, although not explicit in program goals, composites programs are desired by the chemical companies which provide the materials that go into composites. Ciba-Geigy and DuPont, for example, provide free materials to the Sailplane Project, in return for complete access to the results of various tests.

The third issue is manpower or personnel. The goal is to provide students with skills and industry with a resource. Special projects such as the sailplane are seen as necessary to attract students, suggesting that the general student market mechanisms have not been sufficient for the goals of the CML. The director mentioned that "we needed a project attractive to undergraduate students. They have no appreciation for the materials discipline." The Institute is interested in marketing both materials: composites and students. A project administrator mentioned the need to create the need for the engineers who are produced through the program.

For the students involved, the Sailplane project provides "hands-on" experience, a chance to see if the classroom experiences are relevant to the "real world," and opportunities to gain experience for future employment. A large number of the participants are ROTC students, and posters of fighter aircraft and military vehicles in the workspace are additional indicators of anticipatory socialization at work.

Sponsorship for the Sailplane project has come predominantly from NASA and the ARO, and they report receiving information about novel composites fabrication processes and a new configuration of composites for particular kinds of structures as a result of their interactions with the Institute. Earlier work in aircraft components was apparently well received by commercial aircraft sponsors. With the complexity of the current sailplane and anticipated future projects, the benefits are less clear cut. Insistence on "radical design" with self-powered ability makes it very difficult to sort out the benefits of the interactions of composites, structures, and manufacturing techniques. This complexity is necessary to maintain sufficient intellectual rigor in the projects so that they are more than assembly-line manufacture and testing work and have an educational mission. However, this pushes the projects beyond a reasonable time-frame for completion, and delays in the project reduce many students' participation to manual labor rather than design. One product is well received: a small but steady supply of composites-capable students.

## THE METROLOGY LAB: ARCHITECTURE OF AN AGREEMENT

This story begins with the president of an electronics engineering company (hereafter, "C") expressing interest in the possibility of the Center for Manufacturing

Productivity and Technology Transfer (CMP or CMPTT) performing some research and development for the company in the area of sensor technology. We trace the forging of the relationship and interrogate some different ways to interpret the resulting contract and practices.

According to an interoffice Center memo, the company was founded in the 1930s. It originally provided installation and maintenance service to companies involved in the processing of sanitary fluids (e.g., breweries, beverage manufacturers and bottlers, dairies, pharmaceutical manufacturers, etc.). The company has grown to include the design, development, manufacturing, sales and marketing, installation, and servicing of devices and components and full systems for measuring and monitoring sanitary fluid levels. To do this, it primarily tracks pressure within the relevant parts of a system. The company had ninety-four employees, and sales were estimated to be between $10 and $20 million dollars per year. From the perspective of Center contracts, the company appears to have a "reactive" mode of operation, that is, it provides what customers request. Actors saw a "push" potential. If the company had greater technical understanding of and confidence in their more sophisticated (and unique) sensors, they could "push" their product into a wider marketplace, rather than hoping new products are "pulled" by consumer requests. With this push, the firm could grow significantly.

Two Center representatives met with the president, chief engineer, and national sales manager on January 9, 1987. The president and chief engineer prepared a list of specifications for the desired sensor. The president believes in "buying available technology, rather than trying to re-invent the wheel." He understands how the CMP sees its role and "knows (and accepts) that the kind of develoment he is seeking will cost in the neighborhood of $100K, envisions RPI and the CMP as a complement to his company's own engineering capability, and seems anxious to proceed." He also undersands that the New York Science and Technology Foundation could be a resource for possible assistance, but is generally "not interested in depending on the government."

As a result of this meeting, the CMP representatives determined that "given a sound technical proposal and a reasonable cost (estimated at $75K to $100K), C would fund a project at the CMP," and "there is also a high probability the CMP could develop a sensor with the capability C is seeking." The next set of questions that had to be answered were: (1.) Can the development of such a sensor be accomplished at all, and if so, with a very high probability of success, so as not to waste C's money? (2.) Can it be developed in an acceptable time period (nine to twelve months) . . . to satisfy market demands. (3.) Is it appropriate for a student project? (4.) Can it be made 'affordable' ($75–100,000)? If the answers to these questions were positive, there would be a follow-up meeting with C at which a detailed statement of work would be developed. This would be followed by the preparation of a formal technical and cost proposal, targeted for a June 1 (1987) "go ahead." For this start date, the final proposal should have been submitted by March 1 and project approval secured.

A preliminary proposal was sent to the president of C on January 28, 1987. He had several questions which are answered in a letter to him dated February 11, 1987. He apparently had a question about the Project Agreement attached to the preliminary proposal and was told that this is "typical of legal agreements that RPI and the CMP use." In this letter the estimated project cost is broken down as follows:

| | |
|---|---|
| Personnel | |
| 2 seniors | |
| 1 graduate | $69,000 |
| Materials/Fabrication | $15,600 |
| Tuition for Graduate | $5,400 |
| Total: | $90,000 |

All dollar amounts included Institute overhead and represented the approximate direct cost to C. The final proposal was sent to C on May 11, 1987.

The significance of this contract is indicated in a memo from one of the Center directors to the provost. In this memo, the director lists corporations that have signed on for research project efforts or memberships in the CMP (including GM, IBM, GE, Eastman Kodak, ALCOA, etc.). The director then notes: "While these include technological leaders in aerospace, electronics, automotive, primary metals, and consumer industries, they do not include any of the companies that comprise the 'backbone' of America, specifically organizations that do $50M/year or less in gross sales."

The participation of smaller companies is viewed as consistent with Rensselaer's goals. The link is that RPI has "resources and expertise that these companies cannot maintain in-house, and they have positions for our engineering graduates." The competitive gains to these companies, which would probably be local, would be advantageous to both RPI and New York State. RPI's commitment to "growing local industry" includes an Incubator program and a Technology Park. New York State has made a "substantial investment in research for industry through its Science and Technology Programs at the CII." In this memo, a proposal follows to change the overhead structure. In particular, an overhead discount is recommended. The current CMP surcharge on all non-member project costs is "50.7 percent of the total burdened project price. [RPI standard overhead charge on fixed-price contracts.] The surcharge is dropped to 35.7 for industries with less than $50M/year gross sales. It is suggested that RPI make a proportional adjustment in its overhead charge, dropping the rate from 60.7 to 48.7 [percent]."

The written agreement between the CMP and C was formalized as follows. C has a problem and invites representatives of the CMP to the company to explain that problem. C wants a device called a pressure sensor to broaden the company's technology base and offer a more competitive product. The CMP says that it could design a pressure sensor for C. Their interest in this project is that it will give them an opportunity to solve an engineering design problem consistent with their educational and research goals. These precontractual expressions of reciprocity appear in a letter from a Senior Project Manager (SPM) and a Project Manager (PM) of the CMP to a C officer.

The agreement is made concrete with a very specific contract. The CMP Project Managers send C a proposal titled "High Performance Pressure/Level Sensor." The proposal is organized into the following sections and subsections:

Section 1, under the names of the SPM and PM includes: Task, Pressure/Level Sensor Specifications, Team, Approach, Obligations and Deliverables, A Time Table. Section 2, under the name of the CMP and the Institute, is titled "Project Agreement" and is divided into the following sections: Project, Performance, Obligations of Client, Changes, Confidentiality, Rights to Patent and Copyright, Independent Contractor, Agreement with Others, Cancellation, Governing Law, and Notices. There is then a written and signed agreement, a document. This document identifies itself as the *Entire Agreement*: "This agreement, all Schedules, and any proposal attached hereto embody the entire understanding and contractual agreement between RPI and Client."

The implication of this contract is that the signers are bound by part IX of section two, "Governing Law." This part articulates that the agreement is governed by the state, that the two parties accept the legitimacy of the state, its courts, laws, and power to use the means at its disposal to settle or attempt to settle any disagreements between RPI and Client. It outlines what the industrial client want, and what the CMP proposes to do, who will carry out the project, and how this will be done. The client wants a pressure sensor (PS), with specified performance capabilities (1, 2, and 3 in the contract), and with specific mechanical (1a, b, c, and d), and electrical (2a, b, c, and d) properties. Ideally they want to produce these devices for a cost of $200–300 per unit, in lot sizes of 100 units at 1000 units per year. It is also desired that this device be manufacturable within the client's current facilities.

The CMP specifies its task by identifying the amount of time and dollareffort they estimate they will need to produce concepts for a PS that meet the client's specifications and to turn the concepts into proof-of-concept prototypes. The client wants a PS with improved performance characteristics. The proof-of-concept prototypes will be evaluated in terms of their actual versus predicted performance. The contract goes on to state the final report will be delivered to the client detailing the design of the prototypes and the results of the performance evaluations. The contract specifies that the project will be carried out by two co-principal investigators (members of the RPI faculty) and a graduate student in Electrical and Computer Systems Engineering who will be working "to complete research requirements toward an MS."

The contract lays out the nature and stage of the Center's "obligations and deliverables" as well as expectations for the Clients. The Center will, in sequence, produce and deliver a Concept Development Report (CDR), Prototype Design Report (PDR), Proof of Concept Prototype(s), and a Final Report. The client agrees to provide technical representatives, to review the CDR and PDR, and to assist in fabrication, testing, and assembling the prototype(s). The project agreement indicates the educational context. This is one of the ways of identifying and protecting the institutional boundaries of RPI (in terms of real and ideal values). According to the agreement, the CMPTT has been established for the purpose of providing engineering students with a "hands-on" practical educational experience, including the development and application of new technologies. The Client, according to the agreement, has a project suitable for placement in the Center "to be completed by a student team" under appropriate direction by a project manager and faculty members. Items of special interest include statements that indemnify RPI, outline norms of confidentiality, specify the rights of the student to publish academic papers based on the project experience, and statements regarding patent ownership. These specifications are designed to articulate a boundary between RPI and the Client, and between education and industry.

The Concept and Design Reports were written in July and October of 1987. These two documents were prepared prior to our fieldwork. The concept report was manufactured out of references and a first order engineering analysis. Textbooks, such as those written by Giovanni and Timoshenko, were used extensively. For example, the report states that "The solution taken for Timoshenko is consistent with that of Giovanni only for the values of q<3 (i.e. H/h < 2.3). The remainder of this performance evaluation will rely on the solution given by Giovanni." Other resources used to assemble the report were descriptions of manufactured items and manufacturer's specifications as well as an industrial standard STD bus, an IBM-PC compatible INAZ 80188 CPU, and related objects. These components are already existing, well characterized, and mostly commercially available products. A test strategy, for the assessment of the prototypes, was also incorporated into the report.

The deliverables to C at the end of the project fell short of fulfilling all of the original goals of the contract between C and RPI. In the end, the company still had work to do to meet its objectives for a fully operational sensor. One of the reasons for this was the conflict between a contract in the corporate sense and the nature of the research and education framework within which graduate degrees are pursued. The letter of the contract was met in this case without fully achieving C's corporate goals. This small episode captures some of the large textures of the emerging changes in the business of technoscience. Our work revealed some of the small-scale features precipitated out of late twentieth century changes in global political economy. Less grandly, we see in the stories we have told signs of fundamental changes in the how, where, and why of work, science, and technology in the contemporary dynamics of politics and economics on the world scale.

## CONCLUSION

Laboratory experiences are socialization processes for students as well as processes which produce new knowledge. The socialization of engineering students toward industrial interests has been of historical as well as anthropological interest (Noble 1977). These stories about laboratory construction, sailplanes, and contracts begin to tell us about what is expected of students, particularly as products or outputs of university activities. Much of the material here indicates the socialization of students to accept a patronage system of directed research. Knowledge is certainly a product, and "learning is the new form of labor" (Zuboff 1988, 395). As reported to DARPA and ONR representatives, "Our main product is students . . . engineers; students who think of composites as they would of conventional materials." Students are not only learning intellectual topics or manual skills, they are learning interaction styles and norms of academic work. Whether graduate students in the DARPA/ONR facilities or students working on the sailplane, there are ways of doing things, ways of belonging, and desirable social connections to be made.

Students are not the only medium for moving knowledge and information around. Faculty view knowledge as something whose value is derived in use. As a member of the faculty remarked to ONR representatives during a site visit, "It is no good if research sits here and nobody capitalizes on it." Research transfer issues are reiterated in annual reports and memoranda, and are linked to goal-oriented, and especially marketable, research products. The annual reports to DARPA and ONR indicate that while commercial, nonmilitary involvement is highly desired, these agencies continue to have a strong influence in determining the research activities in the lab. At the 1989 annual presentations, the materials vendors and industrial actors were described as "difficult" and unreliable to work with. As noted above, not only must students be produced, but the need for the students must be established.[8]

The account of the laboratory construction process is filled with complicated negotiations and expressions of norms and values. This facility will shape (has shaped) the research agenda of the CML for a long time, in fact decades, directing student research, shaping interactions among students and faculty, and providing a locus for group identification as lab and program. With the facility linkages are built and broken to other laboratories and more importantly to sponsors. There is a strong loyalty or at least a resource dependency that inspires what passes for loyalty, and this is conveyed to students with great efficiency.

A research center is a place for the development of links to industrial interests. But center activities potentially have influence well beyond the center's boundaries, and sometimes explicitly so. For example, the incorporation of composites courses into the curriculum, and the specific messages to students and about students as products, illustrates that boundaries among the various institutions are porous.

Sometimes this porosity is troubling. In the case of the Sailplane, framing extensive materials testing for industrial sponsors as an educational goal and benefit to students is one such blurring of boundaries. Other issues of interest to those who study UIRRs include secrecy. A faculty member reported to DARPA representatives that companies were unwilling to participate because of the patenting and disclosure policies at the university. Another noted that work was done for a sponsor on the condition that none of the results were to be disclosed, and the sponsor requested that all unused materials be returned. Elsewhere in the materials surrounding the DARPA project, patenting conflicts are cited. Foreign students are a security issue for DARPA, especially as self-actualized "products" who might return to their home countries and take knowledge with them. The metrology laboratory and contract had a different locus of secrecy issues. Rather than being about national military security, the question was about the security of intellectual property in a competitive marketplace.

The metrology contract and laboratory also draw attention to a contemporary social phenomenon of immense significance, that is, the rationalization (or reduction, or degradation) of mental labor. Other analysts have already noticed the rationale for the theoretical argument "that intellectual labor is being subjected to the same process of rationalization and control that affected manual labor during the industrial revolution" (Perolle 1986, 111). According to this theory, we should be alert to deskilling and proletarianization in intellectual labor, and more generally to a decline of the middle class (Cooley 1980; Wright and Singelman 1982; Derber 1982; Salaman 1983; Kuttner 1983). These processes are likely, if we agree with Braverman's (1974) argument that "a devaluation of intellectual work to reduce the costs and power of labor is in the interests of business and industrial management" (Perolle 1986, 112).

The rationalization of intellectual labor was a significant reality for Max Weber,[9] and while Marx focuses on the means of production and its relationships to expropriation, Weber (1946, 112) draws attention to the "means of administration": "No single offical personally owns the money he pays out, or the buildings, stores, tools, and war machines he controls. In the contemporary 'state'—and this is essential for the concept of the state- the 'separation' of the administrative officials, and of the workers from the material means of administration is completed" (82). The newest twist in this process is the emergence of a generic administrator capable of operating in any context since all contexts of work have been, or are being, rationalized. All work roles in the division of labor are increasingly being reduced to a common denominator, stripped of the trappings of status, power, authority, and class. Everyone who sells his or her labor is a worker and all workers are equivalent, and so they can all be administered using a general algorithm. Fligstein (1987) outlines a somewhat different mechanism in his account of the rise of finance professionals to corporate leadership: "The basis for organizational power must rest on a claim to solve important organizational problems, and the claim must rest on a form of dependency relationship. Whether these problems are real or percieved as part of the organizational culture or cultural environment is irrelevant" (45). Hence the circulation of university administrators within and among institutions, the increasing isomorphism of assessment and accounting practices, and the emergence of administrative units of professionals who manage the activities of faculty, such as technology transfer officers.

The concepts of reduction and rationalization alert us to the increasing subjugation of professionals and professionals-to-be to "the control and direction of management" (Derber 1982, 8). The novelty of industrial engagements with universities today, and especially those specializing in science, engineering, and technology, is the replication of industrial managerial roles and techniques to such an extent that we can speak of the industrialization not only of the universities but of laboratories, classrooms, research, and teaching. Whatever degrees of freedom were allowed in earlier periods are challenged by the interests in gaining greater control over the social production of ideas and the social production of scientists and engineers.

What empirical indicators do we look for to confirm these conjectures? First, are our workers becoming "detail" workers, unable to choose their own projects or tasks and forced to work at the rhythms and procedures institutionalized in the job descriptions and standard operating procedures of the organization? Second, is their knowledge so specialized and narrow that it no longer serves as a base of power and control in the organization? Third, does the reduction of mental labor mean making machines and thinkers more and more alike (or equivalent in the eyes of the generic administrator).

It is clear that the ready-made distinctions between "academic" or "university" and governmental/military, and industrial/corporate values and processes are being challenged in the Center environments we studied. Organizational as well as value boundaries are being challenged and renegotiated. These boundaries have never separated pure organizational entitites. But we need to be alert to the potential and actual changes in the boundaries and values of these key institutions. Our expectation is that generic adminstrators and their managerial strategies serve as the conduits for the movement of values and organizational forms from the more resource-rich environments of government and industry to the more dependent environments of the universities. This does not only mean that universities will be more and more likely to take on the values and organizational forms of government and industry (as those two institutions converge structurally and culturally), it also means that universities will be increasingly likely to be fully absorbed into the structures of industry. That is, they will be seen, and see themselves, as annexes, to the extent that they can sustain their material identities. They might also become obsolete as their structures and functions become increasingly redefined and molded to the interests and needs of government, the military, and the worlds of industry and finance.

## Abbreviations

| | |
|---|---|
| ACL | Advanced Composites Laboratory |
| AFOSR | Air Force Office of Scientific Research |
| ARO | Army Research Office |
| CII | Center for Industrial Innovation |
| CML | Composite Materials Laboratory |
| (C)CMS | (Center for) Composite Materials and Structures |
| CMPTT | Center for Manufacturing Productivity and Technology Transfer |
| CVD | Chemical Vapor Deposition Laboratory |
| DARPA | Defense Advanced Research Projects Agency |
| EVS | Ethics and Values Studies Program (NSF) |
| HiTASC | High Temperature Advanced Structural Composites |
| JEC | Jonsson Engineering Center |
| MRC | Materials Research Center |
| NASA | National Aeronautics and Space Administration |
| NSF | National Science Foundation |
| ONR | Office of Naval Research |

## NOTES

This study was sponsored by the Ethics and Values Studies (EVS) program of the National Science Foundation (NSF). We wish to acknowledge the contributions of Juan Lucena and Mark Gaylo to this project while they were graduate students in the Department of Science and Technology Studies at Rensselaer.

1. Versions of *Building Labs and Building Lives* were presented at the "Universities and the Global Knowledge Economy: A Triple Helix of University-Industry-Government

Relations," January 3–6, 1996, Amsterdam, The Netherlands, and at "The Sociology of Science: Ethical and Normative Perspectives," May 4–18, 1990, Inter-University Center, Dubrovnik, Croatia/Yugoslavia.

2. See, for other examples, Zenzen and Restivo 1982; Dubinskas 1988; Traweek 1989; Latour and Woolgar 1986; Kunda 1992; Bucciarelli 1994; Downey 1998; Kleinman 1998 and this volume; Knorr-Cetina 1999.

3. Although on a different scale, this is similar to the exploration by the late Diana Forsythe (1996).

4. Newcomer (this volume) discusses the ambiguity of the notion of 'functional' in terms of differing actor's needs and perceptions. This is of course an extension of Pinch and Bijker (1987 ) and especially Bijker (1995).

5. See Gieryn (1999) for a discussion of boundaries and boundary work in science. Newcomer (this volume) also discusses the notion of 'boundary objects'.

6. As noted elsewhere in this volume, Noble (1977) argues that there has always been some degree of interpenetration of science, technology, and commerce in the American university system; what is new is the degree, and the degree to which this is explicit and legitimate.

7. That is, when interpreting their design experience in relation to what Bucciarelli (1994) notes as the mythic design process, they did not challenge the myth, only discredited their design experiences as problematic.

8. This process of creating a need for knowledge or products through students deserves further scrutiny. This probably applies very well to the computerization of higher education. It is not clear what the benefits of computerization in higher education are, but by keeping universities supplied with high-end equipment hardware, and software, vendors create a steady supply of computer-capable students. These students enter the workforce and bring with them the expectation that computers will be in their workplaces, increasing the demand for newer systems.

9. See also, for example, Giddens (1971): "The overall trend towards rationalization in the West is the result of the interplay of numerous factors, although the extension of the capitalist market has been the dominant impetus" (183).

7

# Industry, Academe, and the Values
# of Undergraduate Engineers

EDWARD J. HACKETT, JENNIFER L. CROISSANT,
AND BLAIR SCHNEIDER

## INTRODUCTION

This paper examines the influence of two novel instructional programs on the educational experiences, job values, and life objectives of undergraduate engineering students. Much has been written about the values of working engineers (Kornhauser 1963; Layton 1971; Bailyn 1980; Zussman 1985) and about the effects of undergraduate education on students' values (e.g., Astin 1977, chapter II; Weidman 1989). But little attention has been given to the role of undergraduate experiences in shaping the values and objectives of engineers (for exceptions, see Eichhorn 1969; Gardner and Broadus 1990). Yet engineers are a powerful and influential professional group in the United States, and the undergraduate years are a formative period in which engineering students acquire the technical skills and social outlook of their profession. Indeed, recent innovations in undergraduate engineering education that attempt to interest students in the "real world" problems of industry are predicated on the assumption that both instruction and professional socialization take place during this period. Using data from a questionnaire survey of undergraduate engineering students, we ask how two programs—a cooperative education program ("coop" hereafter) that offers students industrial work experience, and an undergraduate research program ("research") that offers experience in academic engineering research—influence the skills, objectives, and values of participants.

## INNOVATIONS IN ENGINEERING EDUCATION

Undergraduate engineering education is caught in a longstanding "tug-of-war between industry's focus on immediately applicable skills and the university's commitment to fundamental knowledge and understanding" (Office of Technology Assessment 1988). These divergent interests are manifest in two relatively new educational programs implemented at a college we shall call "Hilltop Tech." One innovation is the cooperative education program, which offers engineering students a glimpse of the "real world" of engineering through placement in paying industrial jobs for a period of roughly six months. Most participants are employed full-time in organizations nationwide, but a few work part-time in local companies while continuing their course work. Students derive several benefits from coop positions, including:

1. Work experience
2. An opportunity to apply academic knowledge and skills to practical problems, seeing the tangible effects of their studies
3. A chance to sample and impress potential employers
4. Good pay–more than $1500 per month on average—to defray college expenses

Companies benefit from coop placements in two main ways: they are given ready access to talented, inexpensive, motivated workers, and they are afforded a risk-free look at potential employees without the expense and red tape of formal hiring (and the disruption of layoffs or firing).[1] The national economy, too, will presumably benefit as these young engineers direct their attention to the critical technical problems of American industry. Amid these benefits are some potential costs as well. Chief among them is the possibility that coop participants will acquire the values and perspectives of industry early in their careers, concentrating on matters that influence corporate profits rather than the intellectual problems posed by their discipline.

In contrast to the coop program and its focus on industry, the undergraduate research program allows students to take part in a research project on campus under the supervision of a faculty member. These research experiences may last for a summer, a semester, or longer, and they may provide pay or course credit (but not both at once). Some research positions are located within research centers, including industrially-sponsored centers, while others are affiliated with traditional faculty research projects in departmental laboratories.

The program allows students to experience academic engineering research and to sample the uncertainty and excitement of laboratory investigation. Among the other putative benefits of the research program are:

1. Hands-on experience in the production of new knowledge, putting laboratory skills to work on problems posed by an academic discipline rather than contrived exercises
2. A closer appreciation of academic work and the attractions of the academic career
3. The chance to know and become better known by a member of the faculty

Just as the coop program might convey the perspectives and behaviors of the industry, the research program might communicate academic values and perspectives.

## CONCEPTUAL FRAMEWORK AND ANALYSIS PLAN

We view the research and coop programs as elements of the college experience with potential to influence students' cognitive and social development. For the purposes of this study, cognitive effects include an array of skills and abilities, while social development is represented by a set of life objectives and values.

Drawing on other studies of technical work (e.g., Bailyn 1980; Eichhorn 1969; Zussman 1984), we examined a range of cognitive outcomes, broadly conceived to include general intellectual power, problem-solving ability, leadership, communication skills, knowledge of the profession, and ability to do research (see Measures below, and Appendix for specific items). This variety of outcomes is necessary to capture the spectrum of skills comprised by contemporary engineering and fairly reflects the likely educational effects of the programs.

Social development, here represented by eight life objectives and values, can be influenced by many factors (see, Anderson 1985; Astin 1977; Weidman 1989). John Weidman (1989) has drawn the extensive literature on undergraduate socialization into a "conceptual model of undergraduate socialization" in which lifestyle preferences and values (the ultimate outcomes of concern in this paper) are shaped by students' background characteristics, parental socialization, noncollege reference groups, and academic and social aspects of the collegiate experience. While Weidman's model (replete with multiple, reciprocal causes) could not be estimated directly, it does sharpen our central research question by identifying categories of variables that influence values and proposing mechanisms for their action.

Two theoretical issues are of particular importance in this paper. First (in Weidman's words): "What are the various characteristics of higher education institutions as socializing organizations that exert influences on students?" (1989, 297). That is, what are the structural features of college that shape students' experiences? We propose that the coop and research programs may have such formative influence. Second, how strong is the effect of college programs on values in comparison to other influences? That is, are college programs influential when family background, academic performance, and other variables are taken into account? Accordingly, to estimate the effects of program on values, we will control for students' background characteristics (gender, GPA, reasons for attending college), parental socialization (parents' occupation and educational attainment), noncollege reference groups (the influence of parents, family, friends, and personal interests on career choice), and collegiate experience (year and major).

The analysis will be presented in three stages, each concerned with a set of focal questions:

1. Who chooses to participate in coop or research programs? Why do they participate? In particular, are there differences across educational programs in background characteristics, performance, and reasons for attending college?
2. What skills and abilities do students use during the programs, and how does participation influence students' career plans? These analyses are essential to establish differences in program content and to demonstrate program impact on participants. If no program differences are found, or if the programs do not influence participants' plans, then any subsequent program effects must be viewed skeptically.
3. How does program participation affect students' skills, values, and life objectives? Do differences across programs persist when social background and other relevant characteristics are controlled?

## RESEARCH DESIGN

To compare the programs' influences on students, we conducted a questionnaire survey of undergraduate engineers who were enrolled at Hilltop Tech in the Spring 1990 semester. Hilltop is a predominantly scientific and technical college of about 4,400 undergraduates and 1,600 graduate students.

## Sample

Two samples were drawn: (1) a one-third probability sample of engineering students registered in the Spring 1990 semester, and (2) a one-half sample of sophomores, juniors, and seniors who participated in coop during either of the prior two semesters (Fall 1989 and Spring 1989) and were registered in the Spring 1990 semester. The two samples are combined in these analyses, permitting greater precision of estimate for coop participants. Overrepresentation of coop participants would bias estimates of population characteristics, but such projections are not of concern in this paper.

Questionnaires and a follow-up postcard were mailed to 972 students, yielding 436 usable responses, for a response rate of 45 percent. While this is low for most surveys, the population studied is generally considered reluctant to cooperate with any solicitation that even vaguely resembles something "the administration" might want them to do. The response rate is quite good when compared to other surveys of a similar population: a survey of 2,250 Michigan State University engineering students had an 18 percent response rate (Gardner and Broadus 1990). Although the cover letter clearly identified us as academic researchers (not administrators, a distinction that certainly worked to our benefit), many students perhaps judged us guilty by association and declined to participate.

Table 7.1 shows the main characteristics of the study sample. Sample distributions on such variables as student status, proportion female, and GPA were comparable to those for the school as a whole. Table 7.1 also displays the four categories of students to be compared throughout the paper: research participants, coop participants, dual-program participants (that is, those who did *both* coop and research), and nonparticipants (who took part in neither program). Since very few freshmen and sophomores participate in research or coop programs, the analyses that follow will be restricted to the 236 juniors and seniors in the sample.

## Measures

We will be concerned with two main sets of outcome variables: (1) educational outcomes, which include skills and abilities acquired during college, and (2) values, which include desired job attributes and life objectives. The first set of variables measures the improvement in skills and abilities that students attributed to their college experience, while the second reflects students' values for their work and their lives. Let us describe each set in greater detail.

*Educational outcomes*. Respondents were asked to indicate whether their education increased, reduced, or left unchanged each of twenty-one skills and abilities. Through factor analysis these items were combined into five additive scales (one item, ability to do research, remains by itself). The scales are: ANALYSIS (ability to identify, analyze, and solve problems), INTELLECT (general intellectual abilities such as creative thinking, ability to learn, ability to do research, competence in field), COMMUNICATION (e.g., ability to communicate and work with others), PROFESSIONALIZATION (competence, knowledge of field's requirements), and LEADERSHIP (ability and desire to lead others). These items, representing the range of likely consequences of undergraduate education, were adapted from a questionnaire administered in the 1970s to MIT graduates (Bailyn 1980). Constituent items and scale reliabilities are presented in the Appendix.

TABLE 7.1.

## CHARACTERISTICS OF THE SAMPLE STUDENT
## STATUS BY EDUCATIONAL PROGRAM

| | EDUCATIONAL PROGRAM | | | | | |
|---|---|---|---|---|---|---|
| STATUS | RESEARCH | COOP | BOTH | NONPARTICIPANT | TOTAL | % OF ALL |
| Freshman | 6 | 0 | 0 | 103 | 109 | 25 |
| | 6% | 0% | 0% | 94% | 100% | |
| Sophomore | 5 | 5 | 0 | 77 | 87 | 20 |
| | 6% | 6% | 0% | 88% | 100% | |
| Junior | 32 | 29 | 9 | 54 | 124 | 29 |
| | 26% | 23% | 7% | 44% | 100 | |
| Senior | 29 | 40 | 20 | 23 | 112 | 26 |
| | 26% | 36% | 18% | 21% | 101% | |
| Total | 71 | 74 | 29 | 257 | 432 | 100% |
| | 16% | 17% | 7% | 59% | | |

| MAJOR | N | PERCENT |
|---|---|---|
| Electrical, Computer Systems, & Electric Power Engineering | 119 | 30% |
| Mechanical & Aeronautical Engineering | 115 | 29% |
| Chemical & Materials | 62 | 16% |
| Industrial Engineering | 28 | 7% |
| Biomedical, Environmental, & Civil | 52 | 12% |
| Nuclear, Engineering Sci. & Other | 14 | 4% |
| Total declared majors | 391 | 98%* |
| Undeclared or missing | 45 | |
| Total surveys | 436 | |

| SEX | | |
|---|---|---|
| Female | 82 | 19% |
| Male | 354 | 81% |

*Rounding error

*Values.* Eight scales were constructed to capture respondents' life objectives and preferred job attributes. Scales and constituent items were adapted from prior studies of undergraduate values (e.g., Anderson 1985; Astin 1977; Bailyn 1980). The main dimension that underlies all eight scales is the choice between intrinsic and extrinsic rewards. Four specific scales reflect intrinsic values: INTRINSIC (desire for intrinsic rewards from work), ALTRUISM (willingness to help others), POLITICAL (desire to be politically active), and RECOGNITION (desire for recognition). Rewards of an extrinsic character include DIRECT (the desire to have authority for the work of others), SAFETY (importance of safe working conditions), COMFORT (importance of a comfortable life), and EXTRINSIC (desire for extrinsic rewards).[2] Constituent items and scale reliabilities are presented in the Appendix.

## Limitations of the Study

This is a study of one school at one point in time, thus the results can hardly be generalized. Moreover, the cross-sectional design of this study cannot conclusively distinguish between the attitudes and values students bring with them to college and those they acquire on campus. But the study can indicate likely effects, eliminate specific competing explanations and confounding variables, suggest future lines of investigation, and offer an empirical counterweight to speculations about such programs' effects. Indeed, the specter of selectivity bias hangs over any study that does not randomly assign students to programs, because it will always be impossible to exclude interactions among program, maturation, history, unmeasured independent variables (in this case, predisposing attitudes), and outcomes (Cook and Campbell 1979). Thus the one-school, one-time, cross-sectional design, while not optimal, is a useful starting point for this line of inquiry.

## RESULTS

The results will be reported in three main parts, corresponding to the three sets of research questions outlined above. Part one reports differences in background characteristics between program participants and nonparticipants; the second part compares the programmatic experiences of participants; and the third part assesses the effects of participation on educational outcomes and values.

## Participant Differences

Are there differences between program participants and nonparticipants in age, gender, year in school, GPA, social class origins, and reasons for attending college?

TABLE 7.2.

## SOCIAL BACKGROUND CHARACTERISTICS
## BY EDUCATIONAL PROGRAM

|  | RESEARCH | COOP | BOTH | NEITHER | F (SIG)* |
|---|---|---|---|---|---|
| Age (years) | 21.7 | 21.1 | 21.5 | 21.2 | 1.5 (ns) |
| Proportion female | .20 | .22 | .21 | .21 | .0 (ns) |
| Parent occupation | 7.8 | 8.0 | 8.2 | 8.0 | .2 (ns) |
| Parent education | 4.1 | 4.8 | 4.8 | 4.5 | 4.0 (<.01) |
| GPA | 3.2 | 3.1 | 3.2 | 2.9 | 5.2 (<.01) |

* F statistic based on a one-way ANOVA across four categories (df=3, 232).

Differences in social background would suggest selection mechanisms sorting students into different educational programs. To the extent possible such differences must be taken into account when examining program effects and other outcomes.

Participants are about the same age and gender as their counterparts, with somewhat better academic performance as measured by GPA (see table 7.2). The GPA difference is unsurprising as there are both selection and self-selection mechanisms at work here that will draw off the "cream" of the student population into these programs. Not only will employers and professors choose more capable students, but more able students are also more likely to undertake coop or research programs.

More suggestive are differences in the social class backgrounds of students who participate in the coop and research programs. Table 7.2 shows that the social class origins of coop participants are somewhat higher than others', and the class origins of research participants are somewhat lower than others'. Table 7.3 examines the relationship between program and social class origins in greater detail.

Social class origins were defined in two ways: by parents' occupation and by their educational attainment. Occupations were divided into two broad, equal-sized categories: high white collar occupations (professionals and managers of businesses employing 25 or more persons) and others. Students were assigned to the category corresponding to the highest occupation of either parent. Educational background was divided into four categories, based on the highest educational attainment of either parent: high school or less, some college (but not a four-year degree), a bachelor's degree, and any postgraduate degree.

Table 7.3 displays the relationship between social class and the program in two ways: as choice, showing the distribution across programs of students with

TABLE 7.3.

## PROGRAM CHOICE AND COMPOSITION BY PARENTS'
## OCCUPATION AND EDUCATION COMPOSITION
[column percent] in ()

| Occupation | Research | Coop | Both | Neither | Total | N |
|---|---|---|---|---|---|---|
| Father | | | | | | |
| High white collar | 22% | 31% | 11% | 36% | 100% | 112 |
| | (41%) | (51%) | (41%) | (52%) | (48%) | |
| Father | | | | | | |
| Other occupation | 29% | 27% | 14% | 30% | 100% | 124 |
| | (59%) | (49%) | (59%) | (30%) | (52%) | |
| Chi-Squared-2.4, 3d.f., n.s. | | | | | | |
| Either parent | | | | | | |
| High white collar | 23% | 30% | 10% | 37% | 100% | 120 |
| | (46%) | (52%) | (41%) | (57%) | (48%) | |
| Either parent | | | | | | |
| Other occupation | 28% | 28% | 15% | 28% | 99% | 116 |
| | (54%) | (48%) | (59%) | (43%) | (52%) | |
| Chi-square=2.9, 3 d.f., n.s. | | | | | | |
| Highest parental education | | | | | | |
| High school or less | 43% | 17% | 5% | 36% | 101% | 42 |
| | (35%) | (11%) | (8%) | (21%) | (20%) | |
| Some college | 16% | 34% | 6% | 44% | 100% | 32 |
| | (10%) | (18%) | (8%) | (19%) | (15%) | |
| College degree | 21% | 35% | 16% | 28% | 100% | 57 |
| | (23%) | (32%) | (35%) | (22%) | (22%) | |
| Graduate study | 21% | 30% | 16% | 33% | 100% | 81 |
| | (33%) | (39%) | (50%) | (38%) | (38%) | |

Chi-squared=16.6, 9 d.f., p=.05
Hierarchical loglinear models for occupation (0)) by education (E) by program (P)

| Model | Chi-squared | DF | P |
|---|---|---|---|
| 1. O E P (independence) | 85.2 | 24 | .00 |
| 2. O★E O★P (occupation effect) | 29.7 | 18 | .04 |
| 3. O★E E★P (education effect) | 12.6 | 12 | .40 |
| 4. O★E E★P (education/occupational effects) | 6.2 | 9 | .72 |

similar social class origins, and as composition, showing the class characteristics of each program. According to the first row, of all students whose fathers were employed in high white collar occupations, 22 percent participated in research, 31 percent in coop, 11 percent in both programs, and 36 percent in neither program. In contrast, the second row (in parentheses) shows the social class composition of each program. (For example, 46 percent of all research program participants had a parent with high white collar occupations, and 54 percent did not.) The data suggest only a slight tendency for students of lower social origins to participate in the research program: 23 percent of students with a high white collar parent chose to do research, whereas 28 percent of students with parents in lower status occupations chose to do so. The class composition of the programs tells a simpler story: a little more than half of those who did coop alone (or took part in no program) come from high status families whereas less than 46 percent of those who did research have high social origins.

Differences by parents' education are more pronounced and more complicated. Students whose parents had no more than a high school education are two or three times more likely than others to participate in research ($43\%/21\% = 2$; $43\%/16\% = 2.7$) and are about half as likely as others to take a coop position ($17\%/34\% = .5$; $17\%/30\% = .6$). Students whose parents received college degrees are about three times as likely as others to do both research and coop ($16\%/5\%$ and $16\%/6\%$).

The joint influence of occupation and education on program choice was evaluated by fitting a log linear model to the cross-tabulation of education, occupation, and program. As the bottom panel of table 7.3 shows, program choice depends significantly on education and, to a lesser extent, on occupation. Incorporating a direct effect for occupation or for education (and fitting the joint distribution of the independent variables, education and occupation) produces models that fit significantly better than the model of independence (for occupation, for example, compare the chi-squared statistics of model 1 and model 2: $85.2 - 29.7 = 55.5, 6$ df, $p < .005$). Education has a stronger influence on program choice than occupation: model 3 reproduces the original data more accurately than model 2 (which yields predicted cell frequencies that are significantly different from the original data, $p < .04$). Model 4 shows that the influence of occupation on program choice, while more modest than that of education, is nonetheless distinct from it: adding direct effects for both education and occupation (but no three-way interaction among education, occupation, and program) produces a marginally better fit than the model containing education alone (the difference in chi-squared is $6.4, 3$ df, $p = .10$).

The social class background of students, indicated by their parents' educational attainment and occupational status, significantly influences their choice of educational program. The pattern of this relationship seems counterintuitive, for one might expect students from college-educated families to choose the "academic" option—to do research—rather than take a coop position or stick to their books. One reason for this expectation is that such families are probably better off than those who were not college educated, so the financial need to take a coop job would be lower. (Recall that coop jobs paid an average of $1500 per month for a six-month engagement.) Also, college-educated families are more likely to espouse traditional academic values (i.e., scholarship and research), which, in turn, would influence their children.

Instead, the observed results are consistent with a different lesson passed from parents to children. Perhaps college-educated parents realize that a college education alone is not sufficient to ensure a good career and that augmenting a college degree with work experience provides a valuable advantage in the job market. Perhaps also these students are more aware of the educational options available to them and are sufficiently secure about their career plans, academic performance, and financial resources to leave campus for a time, confident they will return and complete their degrees. (Support for this view can be found in the high rate of nonparticipation among students whose parents received some college education (but not a degree): since such parents are more likely than others to have dropped out of college, they may instruct their children to avoid all distractions. Students whose parents had only a high school education may choose to participate in research as a compromise between their financial need for part-time work and their commitment to completing their degrees.) These decisions are consistent with the views of class reproduction theorists (Bourdieu and Passeron 1977; Willis 1977). If participation in the coop program improves participants' career prospects, then students of higher class origins are preserving their privileged positions by taking part in such programs. Also, participation instills or reinforces in participants certain values and perspectives; then program participation also becomes a mechanism of class socialization and class reproduction.

## Differences in Program Content

What skills and abilities do participants acquire and develop in the programs and how were they affected by their experiences? These analyses are intended to determine whether or not the programs were substantial and influential. If the two programs are not different in content, that is, if both require the same skills in the same degree, then any subsequent outcome differences are unlikely to be "true" program effects but may instead reflect selection differences or other distortion. Similarly, asking whether participants were satisfied with their programs and influenced by them provides an indicator of program impact.

Table 7.4 displays participants' ratings of the skills they needed in their positions. The table is limited to respondents who participated in one or both programs. For clarity, it separates single-program participants from dual-program

TABLE 7.4.

IMPORTANCE OF SELECTED ABILITIES IN
RESEARCH AND COOP WORK

"How important were each of the following skills or abilities in your research (coop) work?"

| | SINGLE-PROGRAM Participants | | | DUAL-PROGRAM Participants | | |
|---|---|---|---|---|---|---|
| | Research | Coop | Sig. | Research | Coop | Sig. |
| Creativity | 3.5 | 3.3 | ns | 3.4 | 4.0 | <.01 |
| Reasoning | 3.9 | 4.1 | ns | 3.8 | 4.3 | .06 |
| Org. skills | 3.7 | 3.9 | ns | 3.5 | 4.0 | .04 |
| Technical skills | 3.9 | 3.4 | .01 | 3.7 | 3.9 | ns |
| Writing/speaking | 3.2 | 3.4 | ns | 2.7 | 3.9 | <.01 |
| Work with hands | 3.4 | 2.6 | <.01 | 3.8 | 3.1 | .03 |
| Work with others | 3.3 | 4.3 | <.01 | 3.6 | 4.5 | <.01 |
| Work accurately | 3.9 | 4.1 | ns | 3.9 | 4.1 | ns |
| Work rapidly | 3.1 | 3.4 | ns | 2.8 | 3.8 | <.01 |
| Work independently | 4.3 | 4.2 | ns | 4.3 | 4.4 | ns |
| Work under pressure | 2.9 | 3.3 | .07 | 2.9 | 3.8 | <.01 |
| N | 60 | 69 | | 28 | | |

participants. Among those who worked in only one program, research partici-pants report significantly greater use of technical and manual skills, whereas coop participants report greater need to work with others and to work under pressure. Those who participated in both coop and research indicated more dif-ferences and larger differences between the two programs, as shown by the right-hand portion of the table. Dual-program participants say that coop positions require more creativity, reasoning, organizational and writing skills, cooperation, and ability to work rapidly and under pressure than do research positions. They agree with single-program participants that research posts demanded signifi-cantly greater use of manual skills.

The pattern suggests that coop jobs immerse students in the world of industry in a way that exercises their intellective, interpersonal, and communi-cation skills—they are neither "gofers" nor window dressing but instead are skilled participants in the work. So involved are they in the work that deadlines and pressures are felt acutely. For their part, research positions afford students the academic experience of less pressure, less collaboration, and greater hands-on experience with apparatus and materials. At its best, such an experience might pass on to participants the tactile dimension of engineering, reinforcing the "sentient knowledge" of the profession (Zuboff 1988, 61–70).

Not only do these programs require the use of important skills but, as table 7.5 shows, they also are quite influential. All students assert that their personal interests guide their career choices (and that faculty advice and family friends are far less influential). Students who participated in coop, either by itself or in tandem with research, rate the coop experience as very influential (4.5), a close second to personal interest and far more influential than classroom experience (a distant third at 3.4). Researchers also rate their program second in influence (3.4), but it is much lower than personal interest and only as influential as class-room experience. Indeed, students who participated in both programs judge research to be less influential than classroom experience and family, rating it 2.9 on a 5-point scale.

Both programs required participants to use a variety of higher-order skills, but coop positions seem to require more of most skills than do research posi-tions. And both programs strongly influenced the career choices of participants, with coop programs substantially more influential than research. We conclude that these are substantial and influential programs—that it is plausible to look for deeper effects on participants—and turn next to an examination of effects on values and life objectives.

## Program Effects on Job Values and Life Objectives

To what extent does program participation influence the skills and traits stu-dents acquire during college? How does program participation shape students' job values and life objectives?

Program participants reported no greater gain than did nonparticipants on the array of skills and traits measured in the survey. The largest difference in table 7.6 is in ability to do research (RESEARCH), where participants in the research program (by itself or in combination with coop) reported a much greater increase in their ability to do research than did coops and nonparticipants. The only other statistically significant difference is in PROFESSIONALIZATION, with participants in any program reporting somewhat better knowledge of their field's requirements than did nonparticipants.

But for all other items—intellectual ability, analysis, leadership, and commu-nication—there are no significant differences among the various categories of students. According to students' self-reports, regardless of the programs chosen, everyone comes away with roughly the same level of improvement in this array of skills.

This is a surprising contrast to the strong influence of these programs implied by the preceding tables. Perhaps the programs' impacts are limited to the specific skills and expectations of participants and attenuate when evaluated against the backdrop of the entire undergraduate experience. It is possible that

TABLE 7.5.

## INFLUENCES ON CAREER CHOICE BY
## EDUCATIONAL PROGRAM

|  | RESEARCH | COOP | BOTH | NEITHER | F (SIG.)★ |
|---|---|---|---|---|---|
| Personal Interest | 4.8 | 6.7 | 4.8 | 4.8 | .5 (ns) |
| Research experience | 3.4 | — | 2.9 | — | n/a |
| Coop experience | — | 4.5 | 4.5 | — | n/a |
| Classroom experience | 3.4 | 3.4 | 3.2 | 3.4 | .3 (ns) |
| Family | 3.1 | 3.1 | 2.8 | 3.1 | .4 (ns) |
| Work experience |  |  |  |  |  |
| Before college | 2.9 | 2.3 | 2.5 | 2.9 | 2.7 (.05) |
| Your friends | 2.6 | 2.9 | 3.2 | 2.6 | 2.7 (.05) |
| Faculty advice | 2.5 | 2.1 | 2.1 | 2.1 | 1.9 (ns) |
| Family friends | 2.3 | 2.4 | 2.3 | 2.4 | .2 (ns) |
| N | 60 | 69 | 29 | 77 | 232 |

★ F statistic based on a one-way ANOVA across 4 categories (de=3, 228)

the skills and abilities acquired through research and coop positions also can be acquired through alternate routes in the normal course of college study.

Table 7.7 presents students' mean ratings of eight job values and life objectives, organized roughly from intrinsic to extrinsic. Intrinsic values and objectives are represented by four scales: INTRINSIC (interesting and challenging work), ALTRUISM (willingness to help others), POLITICAL (desire to influence political matters), and RECOGNITION (desire for the acclaim of others, to write original works). Extrinsic values and objectives are represented by DIRECT (desire to direct the work of others), SAFETY (importance of a secure, safe job), COMFORT (desire for a congenial work environment and time for family), and EXTRINSIC (pay and advancement).

There are no significant differences across program categories on any of the four extrinsic values. All students place roughly the same value on leadership, safety, comfort, and material rewards. But there are significant differences on three of the four intrinsic values. Students who did *both* coop and research are significantly higher than all others in the value they place on intrinsic job characteristics. Students who participated in research, either alone or in conjunction with a coop position, value political activity and recognition more highly than others. These results stand out for two reasons. First, the absence of substantial differences in skills and abilities, reported above, suggested that the programs' influence would be limited. Second, the uniformly stronger effects of

TABLE 7.6.

## SKILLS AND TRAITS ACQUIRED DURING COLLEGE BY EDUCATIONAL PROGRAM

*Listed below are several abilities or traits which people may possess in varying degrees. We are interested in your views of how your educational experiences (at this college) increased or decreased these traits for you.

(Response categories: −2 = decreased much … +2 increased much)

| | PROGRAM | | | | |
| SKILL/TRAIT | RESEARCH | COOP | BOTH | NEITHER | F (SIG.)* |
| --- | --- | --- | --- | --- | --- |
| ANALYSIS | 1.6 | 1.5 | 1.5 | 1.5 | .3 (ns) |
| INTELLECT | .7 | .7 | .6 | .5 | 1.5 (ns) |
| COMMUNICATION | .5 | .8 | .5 | .5 | 2.0 (ns) |
| LEADERSHIP | .7 | .9 | .4 | .8 | 1.8 (ns) |
| PROFESSIONALIZATION | 1.7 | 1.6 | 1.7 | 1.5 | 3.2 (.03) |
| RESEARCH | 1.2 | .6 | 1.2 | .5 | 13.2 (<.01) |

* F statistic based on a one-way ANOVA across 4 categories (de=3, 232)

the coop program in the preceding analyses left us unprepared to learn that participation in research influences such life objectives as commitment to political action and desire for the recognition of others.

To determine whether these differences might have resulted from self-selection according to social background and pre-existing value orientation, a set of multiple regression equations was estimated, predicting each outcome measure using program participation and control variables for social class origins, year in school, gender, GPA, major, career influences, and reasons for attending college. The last set of controls is most important, as these indicate whether a student entered college for extrinsic reasons (such as a better job and more pay) or intrinsic reasons (such as to become better educated and learn interesting things). Program was coded as a set of three dummy variables (with nonparticipation serving as the excluded reference category). Table 7.8 shows the results of these regressions.

All zero-order differences reported above remained significant when social background and other characteristics were controlled. This was tested by first entering all control variables, then entering the set of program dummies and testing whether there was a significant increase in the proportion of variance explained. Net of control variables, students who participated in both coop and research were a quarter-point higher than nonparticipants in their rating of the importance of intrinsic job characteristics. Similarly, research participants rated

TABLE 7.7

## JOB VALUES AND LIFE OBJECTIVES
## BY EDUCATIONAL PROGRAM

| | Program | | | | |
|---|---|---|---|---|---|
| Value | Research | Coop | Both | Neither | F (sig.)* |
| INTRINSIC | 4.4 | 4.4 | 4.6 | 4.3 | 3.7 (.01) |
| ALTRUISM | 3.3 | 3.1 | 3.4 | 3.2 | 1.2 (ns) |
| POLITICAL | 3.1 | 2.6 | 3.0 | 2.6 | 3.5 (.02) |
| RECOGNITION | 3.4 | 2.9 | 3.2 | 3.0 | 4.5 |
| (<.01) | | | | | |
| DIRECT OTHERS | 3.6 | 3.4 | 3.3 | 3.4 | .8 (ns) |
| SAFETY | 3.8 | 3.6 | 3.7 | 3.6 | 1.2 (ns) |
| COMFORT | 4.0 | 3.9 | 3.9 | 3.8 | .8 (ns) |
| EXTRINSIC | 4.0 | 3.9 | 3.7 | 3.9 | 1.6 (ns) |
| N | 61 | 69 | 29 | 77 | 236 |

* F statistic based on one-way analysis of variance across 4 categories (df = 3, 232)

political action and recognition a half-point higher than did nonparticipants—a very large difference on a scale with a range of only four points.

### DISCUSSION AND CONCLUSIONS

Undergraduate research and cooperative education programs strongly influence participants' skills, job values, and life objectives, even when social background and other differences have been controlled. The pattern of effects is complicated. Coop positions more strongly influence skills and career decisions than do research positions, perhaps because they are full-time, longer-term immersions in an off-campus workplace. Yet these strong effects diminish when evaluated in the context of the entire undergraduate career. Research programs had modest effects on students' skills and career decisions, but powerfully influenced intrinsic values and life objectives, particularly the "academic" value of peer recognition and desire to influence the political system. Interestingly, students' social class background (as indicated by parents' educational and occupational attainment) significantly influenced program participation.

    Increased student involvement with industry does not appear to elevate students' desire for extrinsic satisfactions at the expense of intrinsic satisfactions. While there are many sound reasons for regarding university–industry relations with skepticism, a simple sort of "brainwashing" of students through industrial work experience is not among them. Participation in an on-campus research

TABLE 7.8.

## NET EFFECTS OF PROGRAM PARTICIPATION
## ON JOB VALUES AND LIFE OBJECTIVES

Entries are mean differences between programs with other variables controlled. The reference category is nonparticipation ("Neither" in preceding tables).

| VALUE | RESEARCH | COOP | BOTH | $F_{INC}$ SIG.)★★ | $R^{2ADJ}$ |
|---|---|---|---|---|---|
| INTRINSIC | .08 | .19 | .26 | 3.4 (O.2) | .26 |
| ALTRUISM | .13 | .01 | .11 | .7 (NS) | .26 |
| POLITICAL | .48 | .09 | .33 | 3.2 (.02) | .12 |
| RECOGNITION | .48 | .06 | .12 | 4.4 (.00) | .21 |
| DIRECT OTHERS | .31 | .15 | .08 | 1.6 (ns) | .17 |
| SAFETY | .21 | .01 | .09 | 1.2 (ns) | .09 |
| COMFORT | .13 | .09 | .02 | .5 (ns) | .00 |
| EXTRINSIC | .06 | -.10 | -.16 | 1.3 (ns) | .26 |

★Controlling for parents' educational attainment and occupational status, gender, year in school, GPA, major educational goals, and career influences. Control variables were coded as follows. Parents' educational attainment: 3 dummy variables reflecting the attainment of the parent with the greater amount of formal education. The reference category is no more than a high school education, with dummy variables representing some college, college graduate, and graduate work. Parents' occupational status: 1 = high white collar (professional or manager of 25 or more persons); 0 = other. Gender: 1 = female, 0 = male. Year in school: 1 = senior, 0 = junior. GPA: interval scale. Major: dummies representing electrical/computer systems, mechanical (versus all other). Educational purposes: 3 interval-scale variables—intrinsic reasons (scale containing 3 items: general education, become cultured, and learn about interesting things; alpha = .64), extrinsic reasons (scale containing 2 items: get a good job and earn more money; alpha = .77), and a single item, "prepare for graduate or professional school." Career influences: 2 interval scales representing strength of family/friends and personal interest.

★★$F_{inc}$ tests whether adding dummy variables for program participation to the regression equation containing control variables significantly increases the proportion of variance explained.

program, while less involving and impactful than coop on virtually all measures, had a strong and consistent effect on students' values.

These results pose several problems for future research. First, it is puzzling that programs as influential as these had no apparent effect on a host of specific skills and traits. This may reflect a weakness of measurement or design in the present study, but it may also point to shortcomings of these particular programs or of engineering education in general. Second, the cross-sectional design of this study cannot decisively determine that program participation shaped student values: history, maturation, and cohort effects, among others, may also be at work (Cook and Campbell 1979). Similarly, this design cannot elucidate how such perspectives and values are acquired, nor can it show whether they persist into the professional career and shape subsequent behavior. While short-term attitudes and values are inherently interesting and important, enduring values and behaviors are arguably more important. Third, educational programs such as research and coop may play a role in the class reproduction process as arenas of increased opportunity for the privileged and mediators in the transmission of cultural capital. This possibility suggests a new avenue for exploring the role of engineering in the U.S. class system.

## NOTES

The authors are very grateful to Juan Lucena and Javier Bustamante-Donas for their assistance in questionnaire development and to Donna Wagner for help with sampling the student population. Susan Eberley Schatzman and Rachelle Hollander made valuable comments on a version of the paper that was presented at the 1990 meetings of the Society for Social Studies of Science. The project was supported by grant no. BBS87-11341 from the National Science Foundation.

1. Access to students is a principal motivation for companies to participate in university-industry research relationships. Examination of program documents, observation of program operations, and interviews with industrial partners, affiliated faculty, and program administrators (conducted as part of a study of university-industry research relations at Hilltop Tech) all testified to the importance of students for such arrangements. Academic administrators and faculty affiliated with these programs are quick to extol the educational benefits of university-industry partnerships. While the potential effects of industrial contact on undergraduate education have been the subject of much speculation (see Fairweather [1989] for an overview of issues), the matter has received little empirical examination (for an exception, see Blumenthal et al. [1986] who examined effects on graduate students).

2. Compare our scales to those used in the Cooperative Institutional Research Program surveys, reported by Astin (1977, 49): altruism, business interests (being very well off financially, similar to EXTRINSIC and JOB COMFORT), status needs (recognition from others, being an authority in one's field; our RECOGNITION scale), keeping abreast of political affairs (POLITICAL) and having administrative responsibility for others (our DIRECT scale).

3. The absence of differences is not an artifact of scale construction: an identical analysis run on original items yielded virtually identical results. The only difference was coop participants' reporting a marginally greater ability to persuade others (F=2.2, p=.09).

# APPENDIX

Survey of Undergraduate Engineers: Explanation of Scales

SKILLS AND ABILITIES

---

ANALYSIS (Analytic ability)
v150 = ability to identify problems
v151 = ability to analyze problems
v152 = ability to solve problems

| MEAN = 1.44 | ST.DEV. = 0.57 | ALPHA = .83 |

---

INTELLECT (General intellectual ability)
v154 = ability to think creatively
v155 = ability to learn new things
v157 = self-insight
v158 = confidence in career choice
v159 = high career aspirations
v160 = positive attitude toward further education
v161 = breadth of perspective
v162 = overall self-confidence

| MEAN = .62 | ST.DEV. = 0.69 | ALPHA = .79 |

---

COMMUNICATION (Communication skills)
v161 = breadth of perspective
v162 = tolerance of others
v163 = ability to work with others
v164 = ability to communicate
v165 = ability to persuade others

| MEAN = .57 | ST.DEV. = 0.68 | ALPHA = .75 |

---

LEADERSHIP (Ability and desire to lead)
v167 = leadership ability
v168 = leadership desire

| MEAN = .64 | ST.DEV. = 0.82 | ALPHA = .71 |

---

PROFESSIONALIZATION (Professional development)
v149 = competence in field
v156 = knowledge of field's requirements

| MEAN = 1.48 | ST.DEV. = 0.58 | ALPHA = .53 |

---

JOB VALUES AND LIFE OBJECTIVES

DIRECT (Want to direct others)
  v022 = Chance to be my own boss
  v029 = opportunity for leadership
  v139 = responsible for work of others
  v144 = be successful in own business
      MEAN = 3.50              ST. DEV. = .083              ALPHA = .70

EXTRINSIC (Want extrinsic rewards)
  v010 = good pay
  v014 = good chances for advancement
  v018 = an appealing lifestyle
  v140 = being well off financially
      MEAN = 4.00              ST.DEV. = 0.67              ALPHA = .73

COMFORT (Job comfort)
  v016 = time for family and personal life
  v023 = reasonable workload
  v024 = friendly co-workers
  v028 = good location
      MEAN = 3.94              ST. DEV. = 0.65              ALPHA = .67

SAFETY (Hygienic and job security measures)
  v012 = job security
  v021 = well-defined responsibilities
  v026 = highly regarded organization
  v027 = good physical working conditions
      MEAN = 3.71              ST.DEV. = 0.56              ALPHA = .69

RECOGNITION (Want recognition for work)
  v135 = become authority in field
  v136 = recognition from colleagues
  v142 = make contribution to science
  v143 = writing original thoughts
      MEAN = 3.19              ST.DEV. = 0.82              ALPHA = .71

POLITICAL (Politically active/interested)
  v137 = influencing the political structure
  v138 = influencing social values
  v148 = keeping up with political affairs
      MEAN = 2.78              ST.DEV. = 0.96              ALPHA = .73

ALTRUISM (Helping others/self-actualization)
v017 = chance to improve the world
v020 = opportunity to help country
v141 = helping others in difficulty
v145 = clean up environment
v146 = develop philosophy of life
v147 = community action project

   MEAN = 3.30          ST. DEV. = 0.78          ALPHA = .77

INTRINSIC (Want intrinsic rewards)
v009 = interesting work
v011 = chance to work creatively
v013 = challenging work
v015 = chance to develop abilities
v025 = sense of accomplishment

   MEAN = 4.33          ST. DEV. = 0.52          ALPHA = .74

8

# Conflicts of Interest and Industry-Funded Research: Chasing Norms for Professional Practice in the Academy

DEBORAH G. JOHNSON

## INTRODUCTION

One of the most alarming (among many) concerns expressed about increasing industrial funding of university research is that such alliances thrust university professors into situations in which they have "conflicts of interest" (Bereano 1986; Kenney 1987; Bowie 1994). Perhaps the most suggestive examples of this are that of a professor who serves both as a teacher-adviser to a student and as an employer of the same student, and that of a professor who is both a member of a departmental tenure committee and owner of a company using university faculty to do research requiring faculty with certain specialties. While these examples involve professors who own their own companies at the same time they are employed at a university, they point to the possibility of more subtle conflicts arising when an ordinary professor receives industrial funding for her research. The funding, or even the promise of such funding, might influence the professor in advising a student about a thesis topic, in voting on the special-ties to be emphasized in hiring a new faculty member, or in choosing not to publish research results because of their value to the funding corporation if kept secret.

During the two years that I was part of a team studying two university-based research centers seeking funding relationships with industry and govern-ment, I observed professors balancing a wide variety of interests (their own and others) with little accountability for how they did so. This raised questions for me about the proper interests of professors, and this, in turn, led me on a search for principles or values or theories to use in justifying norms for professors.

The criticism of industrial funding of academic research as creating conflicts of interest for university professors is important because it suggests that university-industry alliances might be critiqued from the perspective of professional ethics. It suggests that we should look at university professors as professionals and ask how alliances with industry affect professional practice.

In this chapter I examine the role of professor as a professional role.[1] While such an examination might be done empirically, the analysis here is normative and critical. It presumes that the norms of behavior in a profession are not just the prevailing patterns of behavior, but patterns of behavior that can be justified. They can be shown to be necessary for the good at which the profession aims, to protect those who are served by the profession, and so on. At a minimum it must be shown that they are not harmful to those affected. I begin with an analysis of the concept of 'conflict of interest' and move from there to the role of university professor. Only later do I return to industrial funding of university research.

## CONFLICTS OF INTEREST AND PROFESSORS

While accusations of conflict of interest are common, analyses of the notion of 'conflict of interest'—as an immoral or improper situation for a professional to be in, are not so common. The exception to this is found in the law literature, where a good deal of attention has been paid to conflicts of interest that lawyers should avoid. Davis (1982) has taken the core idea in the legal literature and generalized it so that it applies to other professional roles. Davis argues that a person has a conflict of interest if: "a) he is in a relationship with another requiring him to exercise judgment in that other's service, and b) he has an interest tending to interfere with the proper exercise of judgment in that relationship" (p. 21).

Davis spells this out in much more detail, but these two elements are the core of his analysis.[2] We can use these two conditions as the starting place for an analysis of the role of professor, and it will be useful to break down further the first condition. Condition (a) contains three elements: (1) the person is in a relationship; (2) the person exercises judgment in the relationship; and, (3) the judgment is exercised in the service of the other.

The first two elements of Condition (a) apply to professors unequivocally. Professors, as professors, have relationships with others. Typically they have relationships with students, colleagues, university administrators, government funding agencies, companies (for whom they consult), journal editors, and so on.

And professors exercise judgment in these relationships. They evaluate students, make hiring and tenure recommendations for the university, give advice to companies, evaluate proposals for funding agencies, review manuscripts for journal editors and book publishers, and so on.

It is more controversial to claim that professors exercise judgment "in the service of others." According to the traditional ideology of the university, professors act (or should act) as free agents, in the service of no one, or only in the service of science/knowledge. Perhaps, such a case can be made when it comes to the activities of doing research (e.g., running tests, analyzing problems, gathering data), but such a claim seems implausible when we examine the activities of contemporary professors in their relationships with others. When, for example, they advise students, design curriculum and courses, review manuscripts for publishers, review proposals for funding agencies, or provide consulting to companies, professors judge in the service of others. The advice provided to a student is expected to help the student; tenure and hiring recommendations are supposed to further the interests of the department and the university; manuscript reviews are directed at helping a journal select high-quality papers.

Applying Davis's Condition (b) to professors may also be controversial for, at least on the face of it, to assert that professors have interests would seem counter to the traditional conception of university life. Disinterestness is generally considered a fundamental value of the academy (Merton 1973).[3] But *disinterestedness* cannot mean "without interests." Historically the term was used by some, at least, to mean without economic or pecuniary interests, and it referred to the community of scholars rather than to individuals (Haskell 1984).

Professors have both personal and professional interests. They have economic, romantic, and political as well as intellectual interests. They have interests even when it comes to their research. Ashford (1983) identifies five fundamental decisions that must be made during the course of a research project: choice of general category of research; choice of the specific project; decision on manner and methodology of research; choice of method of evaluation; and decision regarding dissemination of research results. He then identifies factors that motivate university researchers in making these choices: genuine interest; availability of funding; desire for future funding; formal status within the university; and academic reputation (Ashford). These factors are, in effect, interests that professors have.

So professors have interests, very real interests. Can these interests interfere with their judgment? A professor can have an investment which might interfere with her judgment on behalf of a corporate client, or she might have a romantic interest (in a student, say) which might interfere with her judgment on behalf of that or another student. Still, to say that professors have personal interests (such as investments or romance) which might interfere with their judgment is to say something that will be true of anyone making judgments. It is quite another thing to say that the professional life of a professor can itself produce interests that interfere with judgment. This is to suggest that there might be something wrong with the internal structure of the professional role. The question is: can the professional interests of a professor interfere with her or his judgment on behalf of a client?

In keeping with the framework of professional ethics, I will henceforth refer to the others on whose behalf professors judge as "clients." This will push out of focus the fact that professors are employees of universities. Professors will be treated as professionals who independently serve a variety of clients, one of which is their home university. This model fits the behavior of the professors that I observed as well as or better than the model of an employed professional.

## MULTIPLICITY OF CLIENTS

Industry-funded research draws our attention to the interests of one client interfering with the interests of another client. A lawyer may find in pursuing the interests of one client that she will have to sue another of her clients. The fact that she is working on behalf of the first client would be understood to create an interest that would interfere with her judgment on behalf of the second client. According to the ABA code, the lawyer must remove herself from the situation. She must inform the second client and cease to represent her.[4]

It is precisely this type of conflict of interest that industrial funding has the potential to create. Professors may take on interests through corporate clients that will interfere with their judgment on behalf of other clients.

Indeed, what is unusual about science and engineering professors at modern research universities is that they serve a multiplicity of *kinds* of clients. That is, professors serve clients with very different types of interests. Students want knowledge and credentials; universities want indirect costs and want their reputations enhanced; corporate funders want cutting-edge knowledge, publishers want good quality manuscripts identified, and so on.

Generally it is believed that the mix of relationships and interests which converge in a university (and in the work life of a professor) are in harmony. We hear the following sorts of statements: "Teaching and research complement one another"; "Consulting opportunities enhance faculty research"; "Externally-funded research is good for the faculty and students involved, as well as the university, and the funding agency"; "The reputation of a faculty member, built up through extensive travel away from the university, enhances the reputation of her home university." We presume that universities (and the role of professor) are structured so as to bring all the individuals and interests operating there into synergistic harmony. Of course, some recognize that these activities have to be carefully balanced to achieve harmony, but for the most part the balancing is left to individual professors. Yet the potential for conflicts of interest among the mix of clients is enormous. A few examples illustrate:

(1) Protecting the interests of one's university can come into conflict with one's responsibilities as a teacher-advisor when one is advising an unusually good student about his or her future, and it looks as if the student's interests would be best served at another university.

(2) Serving the interest of one's discipline may come into conflict with serving as journal editor if the manuscript being reviewed still needs a lot of work but is written by a young faculty member who is up for tenure; the young faculty member is worthy of tenure, but his publication record does not yet show his potential, and getting the manuscript accepted would help make the case.

(3) Serving the interests of one's university can come into conflict with one's responsibilities to one's students and discipline when a corporation offers funding for a project, and one's university has made it clear that it wants to increase the quantity of funded research at the university, but the research project to be funded by the corporation will not further the education of the students to be involved and will not increase knowledge in the professor's field.

In all of these cases, the professor is expected to make a judgment on behalf of one client, but her relationship with a second client and her responsibility to serve the interest of the second client may interfere with the exercise of judgment on behalf of the first client.

## POORLY ARTICULATED NORMS

The fact that professors serve a multiplicity of clients might not alone be worrisome if the profession had well-articulated norms for behavior in conflict situations and more accountability. Kennedy (1986) puts it this way: "faculty work is relatively uncodified; in a sense, universities are societies without rules. They nevertheless perform rather well, but much of what goes on behind the walls is deeply mysterious to those outside. The missing information amounts to a lesson in accountability" (vii).

Norms of professional practice arise from formal rules and informal conventions. The rules and conventions create expectations in those who interact with professionals. You assume that if you go to a lawyer because you are accused of a crime, and, unbeknownst to you, the crime was against a member of the lawyer's family, then the lawyer will refuse to represent you. When you go to a doctor, you assume that what you say will be kept confidential. These expectations arise from formal rules which require (and allow) lawyers and doctors to behave in certain ways.

Informal rules and conventions function in the same way. Customers expect that auto mechanics will not use a customer's car to tow other cars or make deliveries while the customer's car is in the shop awaiting repair or pick-up. When you go to the dentist, you expect that you will be charged the same fee as others receiving comparable service. You expect when you make an appointment with a doctor that while you may have to wait an hour or so, you will not have to wait days in the waiting room before you will be seen.

When it comes to professors, our expectations—the norms we presume—are largely informal and conventional. Codes of conduct, where they exist, are not well known and there are few legal constraints. (See De George [1997] for the AAUP Statement on Professional Ethics.) Student expectations about the education they will receive are shaped in part by what is written in a catalogue or student handbook and what is said at college fairs and orientation. Generally little is said about professors except that they are highly competent. Norms are communicated from experienced to new members of the professoriat in a very uncontrolled way. Kennedy (1997) reports that he conducted a survey of advanced doctoral candidates and the survey revealed "a disturbing uncertainty and confusion about teaching, institutional governance, and other dimensions of an academic career" (vii). Typically if one teaches as a graduate student, one treats students as he or she was treated. One hears about standards only when they are broken, for example, a student is accused of plagiarism, a professor is accused of sexual harassment.[5] If a student does research under a professor, the student learns from observing the professor how to handle issues surrounding the integrity of research results, or how to represent oneself to funding agencies.

Norms and expectations go hand in hand with an ideology. The ideology explains a profession's (or institution's) reason for being, its ends, the goods it seeks, the values it must uphold, and often the best means for achieving its ends. The traditional ideology for professors and universities has been the ivory tower story. This story explained the commitment to knowledge for its own sake and to transmission of knowledge through teaching and publication. According to the story, both of these flourish in an environment which is minimally structured, controlled, and sheltered from many of the mundane concerns that shape other enterprises.

Whether science and engineering were ever pursued in ivory towers, or whether they once were but have undergone radical change, in the twentieth century is an interesting historical question.[6] In either case, the ivory tower ideology seems no longer to apply to the professional life of science and engineering professors.[7] Knowledge for its own sake will not convince a funding agency to fund a project. The project must promise usefulness, at least in the long term, and if not to industry, then to the nation. Without funding, the knowledge is not likely to be created. Transmission of knowledge is also shaped to some extent by ability to pay. Knowledge goes first (and sometimes only) to those who fund the research, to students who can afford to attend research universities, to colleges that can afford expensive journals, and so on.

Therefore, the norms of professional practice for professors are poorly articulated because there are few formal specifications or rules of practice, and the traditional ideological understanding of the profession can no longer be appealed to in order to explain, clarify, or justify norms of behavior.

## CHOICE OF RESEARCH TOPIC: A SCENARIO

In my observations of two academic research centers I was struck by the powerful influence of industrial funding on faculty and student choice of research topic. Many of those involved acknowledge this influence but see nothing improper or problematic in it. Indeed, some will argue that this is as it should be. Others will argue that the influence can be tempered by institutional policies which ensure that faculty initiate research for which corporate funding is sought, rather than vice-versa.

Consider the way graduate students choose thesis or dissertation topics. Suppose a graduate student approaches a professor to discuss her idea for a thesis topic. Having taken a course on X from the professor, the student is excited about the possibility of doing research on X. The professor remembers that the student was one of the best students in the course, and also remembers from her application to the program that she has a very impressive undergraduate record. He spends some time talking to her about X. Then he goes on to explain that he has a large grant to do research on Y. If the student is willing to do research on Y instead of X, he can offer her a research assistantship, and the research done under the assistantship could be written up as a thesis. The professor spends a good deal of time explaining how interesting and important Y is.

On the one hand, it appears that the professor has done nothing wrong. Indeed, what the professor did is standard practice in many places. Still, the practice can be called into question. If the student is being diverted from something she is really interested in to something which will neither result in a good thesis nor further her future in the discipline, then the professor might be seen as violating his responsibility to the student. He might be seen as using the student to further his own career interests. We can make this possibility clearer if we suppose that another professor had a grant to do research on X and was in need of students to work on the grant. Our professor knew this, but refrained from giving the student this information. He might have a variety of reasons for doing this. He wants the best students working on his grant. This ensures that the work will get done, and get done well. Having the best students enhances his reputation, furthers his research agenda, and so on.

Whether we say the professor did something wrong or behaved well, we must assert a norm for advising students in their choice of thesis topic. My point is only that a standard of practice for handling such situations is not well articulated in the profession. How such situations should be handled is not well understood by professors or students. The potential for misunderstanding is great, as student expectations with regard to the advice they receive may be very different from the reality.

## LITTLE ACCOUNTABILITY

At the same time that standards or norms are poorly articulated, professors have little accountability. The day-to-day life of a professor is closer to that of someone in private practice than to someone who works in an organizational setting in which there are expectations of keeping certain hours, frequent evaluations, and close monitoring of work done. To be sure, professors have some accountability. They are accountable to their universities by means of evaluations of their teaching and measurements of their research productivity and service. They are accountable via reports to funding agencies, and they are accountable to corporate clients via the results they produce. But they are not accountable for their ethical behavior (except perhaps when they commit the grossest of infractions). And neither individual professors nor the profession as a whole are accountable to the public.

Another scenario may be helpful here. An editor at a prestigious publishing company receives a textbook manuscript on Z. She calls several individuals who have published books with the company, and she asks for names of scientists who could evaluate the manuscript. She finds out that Z is a relatively new area and is very likely to grow rapidly in the next few years. There is one professor who has published the first textbook in the field and who is considered the best in the field. She calls this professor and asks her to review the manuscript. The editor promises payment and requires that the review be done by a certain date. The professor agrees to review the manuscript.

The professor has a conflict of interest. If the manuscript is good, publication of it would interfere with the sales of her textbook. That is, she has an interest that could interfere with her judgment. In this case her judgment is on behalf of the publishing company. One can argue that there is nothing wrong here since the editor is aware of the conflict, and, hence, she should take this into account when she makes use of the professor's review.

I am not arguing whether or not the practice should go on. My point is only that expectations are unclear. What are the professor's responsibilities to the editor and to the author? Should the professor disclose the conflict of interest?

Should she make an effort herself to be fair to the author? or is it acceptable simply to say the most critical and damaging things she can think of? What should the editor expect? Is she being fair to the author? Suppose we say that the standard among professors is to try to judge in a disinterested way. Where or how is this norm promulgated? How is it communicated to new professors?

Norms are unclear in situations like this, and there is little accountability. Suppose that despite its virtues, the professor harshly criticizes the manuscript and recommends against publication. The author will have difficulty publishing the book if each publisher is directed to our professor, the expert in the field. Suppose, nevertheless, that the author manages to get the manuscript published. Suppose further that the book is a great success. Will there be any repercussions on the professor? Perhaps, in the long run, the editor might see that the professor did not act in the best interests of the company. This, however, is not likely to have any consequences for the professor.

So, it appears that professors, as professors, manage a multiplicity of clients, have poorly articulated norms or standards of behavior for dealing with their clients, and have little accountability. These characteristics seem to create enormous potential for conflicts of interest.

## GUNS FOR HIRE VS. PROFESSIONALS

These characteristics of the role of professors come into focus again when we ask what it means to be a member of a profession. The literature on professions is replete with characteristics of professions (Bayles 1981; Hughes 1963; Greenwood 1957). Becoming a member of a profession typically involves mastering an esoteric body of knowledge. This, in turn, may lead to a division in the profession between those who continue to develop and renew the body of knowledge (researchers) and those who use the body of knowledge (practitioners). Typically there is a higher education requirement; that is, members must master the esoteric body of knowledge at an institution of higher education. Use of the body of knowledge or the skill associated with its use is thought to serve an important social function. Members organize themselves, and through their organization regulate admission to the profession and promulgate standards of practice. Often a code of conduct is created which establishes standards for members and expectations for the public. Society recognizes the profession and its representative organization by granting it the legal right of self-regulation with the understanding that the profession will serve an important social function. This right gives the profession a monopoly on the provision of services in its domain of authority. This often means that the profession will not be subject to public control, as only members have the expertise to judge or evaluate the work of members.

The professoriate possesses some of these characteristics. It serves an important social function (creation and distribution of knowledge) and is not subject to public control. To practice, one must master an esoteric body of knowledge. But while admission is difficult, it is not controlled by a single organization. Indeed, there is no single organization for the profession. Requirements for admission and standards are, to a large extent, locally controlled, in the sense that different disciplines and different universities set their own standards. The professoriate as a whole lacks a code of conduct or any kind of mechanism for setting and monitoring standards.

It is important to emphasize autonomy here. Traditional professions are protected from outside control both as professions and as individual practitioners. The profession as a whole regulates itself by controlling admission and setting standards. Members in private practice treat clients and work on their own without supervision or monitoring. Even those who practice in organizations are afforded independence. Friedson (1971) writes, "when an occupation has become fully professionalized, even if its work characteristically goes on in an organization, management can control the resources connected with work, but cannot control most of what the workers do and how they do it" (22–23).

Now, while professions are often contrasted with nonprofessionalized occupations, it is more enlightening for our purposes to contrast professionals with "guns for hire." Guns for hire will use their knowledge and abilities to do whatever is desired by a buyer, be it a consumer or an employer-manager. Others, outside the occupation, by paying for the guns, determine what gets done, and how it gets done. In contrast, members of a profession are constrained by standards set by their profession. (They constrain themselves.) Professions have internal values rather than letting the external world shape them. We expect professionals not just to take orders but to lead. We expect them not just to give us what we want, but to help us figure out what we should want. For example, we expect the medical profession as a whole not just to respond to sick patients but to define health, to recommend health policies, to inform us about preventative measures, and so on.

In the 1980s and 1990s professors of science and engineering have come to look more like guns for hire than professionals. They have valuable knowledge and skills and they provide these to the highest bidders. One can counter this by saying that professors continue to control how work (i.e., research, knowledge-creation) is done, but the analysis of choice of research topic suggests they are not controlling the directions of research, collectively or individually. Nor are they controlling standards for dealing with students and others.

The National Science Foundation and other government agencies recognize the autonomy of professors as professionals when they use processes that involve professors both in setting research agendas for directorates and when they use peer review in selecting individual projects for funding. Such processes are, however, rarely involved in industry-funded research.

Industrial funding of academic research threatens to erode the profession's commitment and sense of responsibility for the shape and future development of science and engineering. Moreover, this external control opens the door to an undermining of client-professional relationships, especially student-teacher relationships.

Return again to the professor discussing a thesis topic with a student. Suppose the professor is faced with the criticism that he manipulated the student into serving his interests. He is likely to reply by pointing out that the student was free to say "no" to the assistantship. "After all," we can imagine the professor arguing, "the student ought to talk to other professors. There is no coercion. The student is capable of deciding for herself what she wants to do."

This suggests not that we think of professors as professionals but as salespeople making deals. On this line of reasoning, those who deal with professors ought to presume the operating principle in business, *caveat emptor*—"buyer beware." Just as you ought to shop around for the best deal on a car, you ought to shop around for a thesis topic, advisor, and assistantship.

The business model of professor-student relationships does not come to grips with the disparity in expertise in client-professional relationships. Clients come to professionals because they need or want something done but don't have the expertise to know what is needed or what should be done. The differential in knowledge (expertise) leads to special responsibilities for the professional. It means, on the one hand, that we do not want professionals simply to be agents of their clients. We don't expect them simply to implement the decisions of their clients. They identify and recommend courses of action. On the other hand, even though the professional has knowledge that the client does not possess, we do not expect professionals simply to act paternalistically. If we think of the relationship this way, a client would have to turn all decision-making power over to professionals and become wholly at their mercy.[8]

The fiduciary model of client-professional relations recognizes that trust is at the heart of the client-professional relationship. Clients must trust that the professionals have knowledge and will use it on the client's behalf; professionals must trust that clients are giving them accurate information; and so on. But trust only flourishes when both parties have similar expectations about their roles and when the professional's primary stake in the relationship is serving the client.

Contrasting the trust in client-professional relationships with that between friends, Sokolowski (1991) writes the following:

> The relationship between professional and client is a fiduciary relationship. The client trusts the professional and entrusts himself or herself—not just his or her possessions—to the professional. The professional is presented as trustworthy not primarily in the way a friend is found to be faithful, by having proved himself or herself in many situations, but by having been certified as a

professional. There is an elegant anonymity to professional trustworthiness; if I get sick away from home and must go to the emergency room of a hospital, I can in principle trust doctors and nurses I have never met before, I enter into a fiduciary relationship with them because they are presented as members of the medical profession, persons who are certified by the profession and who can, prima facie, be taken as willing to abide by its norms. I do not have exactly the same kind of trust if my car breaks down somewhere away from home; I am delivered over rather to the personal honesty, trustworthiness, and competence of the local mechanic. It is as though I had to find a temporary friend rather than being able to appeal to a professional. (p. 31)

Notice how clear Sokolowski's expectations are about the character of the treatment he will receive from doctors and nurses and how much less he has to go on in his dealings with a local mechanic.

In the case of the student choosing a thesis topic, the fiduciary model would have the student go to the faculty member expecting that the faculty member will serve her interests, not his, would have the faculty act neither self-interestedly nor paternalistically, but would have him listen and offer advice that helps the student make the decision for herself.

External funding threatens to undermine the trust at the heart of student-teacher relationships because it creates interests in professors that can affect their judgment on behalf of students. It means professors have clients whose interests might be served by a student's decision.

Student-teacher relationships are special in that in addition to a disparity in expertise is a disparity in power. The student cannot afford to alienate her professor because of the credentialing power of the professor. This exacerbates the situation, for while clients of doctors or lawyers are free to complain when the fiduciary relationship is violated, students are under more pressure not to do so.

In any case, students are not the only clients whose interests are at odds with industrial funding. The same erosion of trust can occur when we take a professor's department or university as the client and focus on hiring decisions and the choice of specialty. Professors are expected to judge on behalf of their departments. In hiring decisions, they are expected to recommend specialties and persons who will strengthen the department. Indirectly, of course, this serves students. But a professor's interest in expanding research for a corporate client may interfere with her judgment on behalf of the department about what specialization will strengthen the department the most.

## CONCLUSIONS

This analysis leads us back to two claims that I made near the beginning of this chapter when I argued, contrary to the traditional ideology, that professors have interests and judge on behalf of others. The traditional ideology of the university and of professors was an ideology that served the professionalizing function. Professionalization can be understood as an institutional mechanism for protecting a socially important enterprise from the dangers of capitalism (Haskell 1984). The idea of professors sheltered in the ivory tower ensured that new knowledge would be developed and transmitted not for its own sake, but for the good of humanity. Knowledge would be developed and transmitted without the distortions that might result if either public demand or business interests controlled. Professors would tell us what was important to know and would continue to pursue the development and transmission of such knowledge.

One can argue that professionalization never occurred for the professoriate, or that it occurred but was weakened because of such factors as science becoming so expensive, waning support for higher education, and so on. Whatever the case may be, it is clear that industrial funding of university research exploits the weaknesses of the professoriate as a profession. It does this both at the level of the profession as a whole as it threatens to shape the directions of research, rather than allowing the profession (disciplines) to shape these. As well, it threatens individual practice. That is, it threatens to shape the relationships that professors have with others and especially with students. Lest I be accused of making industry "the bad guy" in all of this, let me add that there are many factors which have led to this situation, not the least of which is how expensive research in science and engineering has become. The professoriate can be faulted for not coping better with the renorming of science and engineering as it has become so expensive and recognized by industry as crucial to its success. The professoriate has a responsibility to shape itself—to recommend ways of institutionalizing and funding science and engineering that will ensure that the profession serves its social function.

It would seem that a good deal of attention should to paid to restructuring the role of professors. If this does not happen and professors (and universities) slip further into the role of guns for hire, then at least the public ought to be warned so that it can develop appropriate attitudes and expectations for "dealing" in and with the academy.

## NOTES

The research for this paper was supported by NSF Grant No. BBS8711341.

1. My focus is on professors of science and engineering at research universities. The analysis does not apply to professors who are *not* expected to do research and consulting or to bring in external funding.

2. In his fuller analysis Davis (1982) specifies that

A person P1 has a conflict of interest in role R if, and only if: a) P1 occupies R; b) R requires exercise of (competent) judgment with regard to certain questions; c) A person's occupying R justifies another person relying on the occupant's judgment being exercised in the other's service with regard to Q; d) Person P2 is justified in relying on P1's judgment in R with regard to Q (in part at least) because P1 occupies R; and e) P1 is (actually, latently, or potentially) subject to influences, loyalties, temptations, or other interests tending to make P1's (competent) judgment in R with regard to Q less likely to benefit P2 than P1's occupying R justifies P2 in expecting. (p. 21)

3. Merton (1973) identifies disinterestedness along with universalism, communism, and organized skepticism as fundamental values of science.

4. The constraints on lawyers are even stronger in the sense that they are required to remove themselves even if there is only the appearance of a conflict of interest.

5. Few graduate teaching assistants are told, for example, that they should not date students in their recitation sections.

6. The history of technological universities may well be very different in terms of this question.

7. Slaughter and Rhoades (1990) work on the premise that science is in the process of "renorming."

8. Bayles (1981) identifies a variety of models of client–professional relationship and argues for the fiduciary model.

9

# Your Space or Mine? Organizational Interactions and the Development of a Two-Arm Robotic Testbed

JEFFREY L. NEWCOMER

To further the understanding of how researchers create a plan to develop equipment and techniques and how they use what they are developing as they interact with the various institutions and organizations, I examine the attempts of one group of engineers to develop a technological artifact, concentrating on the human and nonhuman forces that shaped its development and the manner in which the researchers used what they had accomplished in both their interactions with groups outside of the center and their day-to-day research activities.[1] The group that I consider consists of researchers in a Space Robotics Center (SRC) at a medium-sized technical university, and its attempt to develop an autonomous, two-arm robotic platform, or robotic testbed, to simulate space construction using theories of intelligent control.[2] The issue of external support raises certain questions: how did ties to the sponsoring organization affect the form of the SRC robotic testbed; how did the needs of the sponsoring organization mesh with the responsibilities that the SRC researchers had to the university, the academic community, and their own careers; and how did the SRC researchers use the robotic testbed in their interactions with their sponsors and the rest of the engineering research community? By considering use, I intend to explore issues such as how the SRC engineers used the robotic testbed as a tool to obtain resources for the laboratory (money, equipment, new researchers) and what role the robotic testbed played in legitimating the work of the SRC researchers, especially to sponsor personnel.

One thing that makes the SRC case compelling is that the original grant was not renewed upon its expiration; thus the case has a certain amount of closure to it. However, as with record/playback machine tools (Noble 1984), gas refrigerators (Cowan 1985), military aircraft (Law and Callon 1988, 1992), and many other things (Braun 1992), questions about the "technical merit" of the SRC two-arm robotic testbed constituted only one of many factors that led to the withdrawal of support from the project and the center. The robotic testbed was shaped by the interests and experiences of the SRC research and their negotiations with and understandings of the aims of external actors, especially sponsor personnel. At the same time, it served as a tool for the researchers as they attempted to secure further resources, legitimate their work, and retain autonomy over their own research in a strongly interconnected environment. I begin by giving a description of the framework I used in my examination, followed by a description of the formation of the SRC and the shaping of the robotic testbed. I then continue with descriptions of how the SRC researchers used the robotic testbed and conclude with a set of lessons garnered from the experiences of the SRC researchers and how this case can help inform future studies of research laboratories.

## A FRAMEWORK FOR EXAMINING THE SRC RESEARCHERS

I refer to the SRC robotic testbed neither as an instrument, which carries very narrow connotations, nor as a technology, which carries very broad ones (Winner 1977), but as an instrumentality. The term *instrumentality* was suggested by Price (1984) "to carry the general connotation of a laboratory method for doing something to nature or to the data at hand" (13). At the heart of the concept of 'instrumentality' is the interaction of techniques, equipment, and people. The general implication in the term *instrumentality*, as it has been used, is that a portion of an instrumentality is equipment that is specialized and rare, if not unique. Price's use of the term *instrumentality* also implied something that was already in use. Therefore, to expand the concept of 'instrumentalities' to include equipment, techniques, and interpretations that researchers have not yet developed to the point of being useful for exploring issues in new manners, I will specify what I call an operational threshold as some developmental state of an instrumentality at which point it can be used to explore or demonstrate a phenomenon or event that could not be explored without it. Accordingly, I will refer to instrumentalities that have not surpassed their operational threshold as nonoperational. One example of a nonoperational instrumentality could be a high-energy physics detector that can replicate existing results but cannot yet be used to explore beyond that point due to unfinished equipment or unrefined interpretation of its output. By considering the operationality of the robotic testbed, I show how the SRC researchers began to use it long before it was developed to the point of being useful for the experiments they intended to eventually conduct with it.

To explain the role that the robotic testbed played in the SRC, how researchers organized their efforts to develop the robotic testbed, how those efforts related to other research underway at the SRC, and how everything related to the organizations with which the SRC researchers were associated, I invoke the concept of 'boundary objects' developed by Star (1989) and Star and Griesemer (1989) and further developed by Henderson (1991). Star and Griesemer define boundary objects as "objects which are both plastic enough to adapt to local needs and the constraints of several parties employing them, yet robust enough to maintain a common identity across sites" (p. 393). The robotic testbed served as a boundary object both internally and externally for the SRC. The robotic testbed was designed so that SRC faculty researchers and their sponsor would all see it as a useful project. What each actor saw as being useful was necessarily varied, and the robotic testbed retained enough interpretive flexibility during its early development that each actor could justify supporting its continued development. Yet, during most of the life of the project, all of the actors involved agreed that the robotic testbed was a space robotics simulation platform. Of course, this was in a large part based upon different conceptions of what constituted research in space robotics. In this role, the robotic testbed represented the Center and the link with the sponsoring organization and was used to justify funding smaller projects and generate research output. The SRC researchers could not, however, rely on the interpretive flexibility of the robotic testbed indefinitely, especially as its physical form and capabilities were developed. As such, they attempted to find a way to make the project completable. In other words, the SRC researchers had to balance technical possibilities with individual and institutional goals in organizing and undertaking the development of the robotic testbed. Many of the efforts that SRC researchers engaged in to keep the robotic testbed development a do-able (Fujimura 1987) project were efforts in negotiation. Negotiation took place both internally among laboratory members and externally between laboratory members and sponsors, administrators, and the like. What eventually transpired, however, was that the SRC researchers found themselves responsible to various actors and organizations (including themselves) whose best interests were not always compatible.

While the SRC did exist as a distinct entity with its own goals, structure, and administration, the Center was part of an institutional setting that determined much of its form prior to its inception. With the exception of a few staff members, all the SRC researchers belonged to academic departments as well. The academic departments involved had set guidelines to which the SRC researchers had to adhere, such as course requirements and requirements for advancement. Moreover, the School of Engineering, to which all the SRC researchers belonged, had guidelines for interdisciplinary research; the greater academic community had standards for recognition of researchers; researchers and administrators from the sponsoring organization expected certain types of research from the SRC. Furthermore, many of the organizations or institutions that gave the SRC its initial operating conditions were related as well—their interactions with each other and the SRC researchers formed the space in which the SRC was established. This is what Law and Callon (1988, 1989, 1992) referred to as a "negotiation space." The SRC researchers attempted to convince other actors, through negotiation and the use of the robotic testbed as a boundary object, that the research at the SRC was an aid to the attainment of the other actors' goals as well. The robotic testbed was both a product and a tool. As a tool, the robotic testbed was used by the SRC researchers not only to do research, but also to try to secure resources and legitimate their work to the various people they interacted with. As a product, however, the robotic testbed was formed by the researchers in the SRC under more constraints than just institutional bound-aries. Despite their best efforts to establish their own work as space related, the sponsoring organization retained a virtual monopoly of declaring the space-relevance of research.

    Along with remaining accountable to their own goals and the desires of other actors, the SRC researchers also had a fair number of issues over which they could not negotiate directly. These factors formed boundaries within which they had to work. Economic, technical, physical, human resource (per-sonnel and expertise), and temporal factors cannot be considered to be the same as dealing with human actors (a good argument and flashy presentation will not convince a room to become larger or a robot smaller). I refer to the set of all such factors as developmental boundaries. Developmental boundaries are contingent, and they are moveable and malleable. Developmental boundaries are contingent in that they are interrelated (the equipment required to meet technical specifications may exist but be too large or expensive for the labora-tory members to obtain) and may be unforeseen or arise unexpectedly. They are moveable and malleable in that researchers can potentially work around or overcome them (a new grant, a relaxed deadline, or a different technical specifi-cation could make a project do-able again) either through ingenuity or negoti-ation with other actors, but not by negotiating directly with the boundaries themselves.

    As development of the robotic testbed continued and it came to literally embody various agreements and compromises, it became a fairly inflexible entity.[3] The development reached a point at which it was easier for the researchers to alter their future plans than to adapt the robotic testbed to meet different goals.

Langdon Winner (1977) refers to this occurrence as reverse adaptation: "the adjustment of human ends to match the character of available means" (229). From the beginning of its conception, an instrumentality will have a finite amount of flexibility in its form and capabilities (and interpretation). The longer that development and implementation continue, the less flexible for adaptation to new missions and contingencies it is likely to become. Thus, if an instrumentality becomes too rigidly defined to be used on a wide range of problems, laboratory members are likely to alter the projects they wish to undertake so that they can be accomplished using the instrumentality (both to take advantage of its presence and to advertise its capabilities). In this respect, the presence of an instrumentality in a research laboratory can have as much influence on the direction of research there as the absence of one could.

## SHAPING AN INSTRUMENTALITY:
## THE DEVELOPMENT OF THE SRC ROBOTIC TESTBED

The SRC was one of nine Space Engineering Centers (SECs) started with seed money in 1988. The nine SECs had been chosen through competitive review of 115 proposals submitted by researchers at academic institutions throughout the United States. From the sponsor's point of view, the SEC program goals were twofold: to create a new generation of researchers trained and interested in space-related issues and to create more interest in space research in industry by getting the SECs to forge ties with industry in order to survive in the long run. The SEC program had an annual operating budget of around $10 million. The organization that sponsored the SEC program, however, is a large bureaucratic organization, and this was a fairly small program by its standards.

The SRC was created in response to the SEC program notice, but not out of thin air. Almost all of the faculty members involved with the SRC (and therefore their graduate students) were members of an interdisciplinary Robotics and Automation Laboratory (RAL) prior to the formation of the SRC. The projects underway in the RAL were fairly diverse and unconnected, mostly sponsored by small grants from various corporations and the NSF. When the proposal to form the SRC was submitted, the faculty who were involved were not doing space-oriented research. The stated purpose of the RAL was "to promote basic research in the area of intelligent machines, automation, and robotics of the future." Space was mentioned as a location for the use of the techniques and technologies developed at the RAL, as were nuclear power plants and industrial production sites, but there was no work done at the RAL addressing the specific issues of space robotics prior to the formation of the SRC. Actually, none of the 117 technical reports that had been produced by the RAL as of September 1988 had anything explicitly to do with space robotics. Moreover, it is important to remember that just because the RAL faculty added the label "space exploration" to their work, they did not divorce themselves from their past experiences or interests.

The purpose of the SRC stated in the original proposal differed significantly from that of the RAL only in the area of application. The dual stated purposes of the SRC were:

> To accelerate the development and advancement of the mathematical theory of Intelligent Control Systems essential for the creation of the necessary algorithms for machine intelligence.
> To develop laboratory demonstrations of intelligent autonomous robotic systems with two or more arms, mobile or static, and with sensory feedback. The demonstrations will emphasize the concepts and theory essential to the development of a system for assembly, disassembly, and repair in space. This may serve as a case study for the design of future space exploration systems.

The laboratory demonstrations mentioned in the latter half of the statement of purpose became the two-arm robotic platform, the instrumentality that the SRC researchers attempted to develop.

The initial description of what the testbed would be like was quite brief. According to the original proposal, the testbed would be "comprised of two industrial manipulators equipped with force sensors and a 3D visions system. Both arms will be mounted on a moving platform which simulates the uncertainty and disturbances that will be experienced by autonomous machines in space." The only concrete pieces of information given regarded the robot arms— industrial—and the need for the platform on which the arms were mounted to react as if it were in space. Descriptions of other aspects of the project, such as grippers and a vision system, were minimal or nonexistent. Industrial robot arms were chosen because they were already in the lab and therefore familiar. While familiarity was an advantage, this gave the robotic testbed a standard manufacturing form (and look), so it was difficult for outsiders to visualize it as a space robot. The platform was what would have made the whole robotic testbed space-like.[4] The SRC researchers required a platform with six degree-of-freedom motion capability so that the platform could be controlled to react to motions of the robot arms on it as if it were in space. Overall, the robotic testbed that the SRC researchers proposed to build was a fairly complex system akin to a flight simulator.

## Developmental Boundaries Revealed and Constructed

The SRC researchers were notified that they would receive an SEC grant in mid-May 1988, and funding began on July 1 of that year. It is unclear exactly when work began on the robotic testbed development, but judging from the SRC status report covering the first six months of the Center's operations, the

team members had not carefully investigated the feasibility of what they proposed to develop. Problems arose with the platform for the robotic testbed.
The SRC researchers discovered constraints regarding cost, size, performance
requirements, and delivery time related to the platform for the robotic testbed.
The researchers had planned to purchase a platform, but when they began
shopping, they discovered that what they wanted did not exist.

The SRC specifications called for a platform with six degree-of-freedom
motion capabilities that had a motion response time at least as fast as the end
effectors on the robot arms (so that the platform could accurately mimic one
floating in space), and also fit into a relatively small laboratory, yet was large
enough for two robot arms to be mounted onto it and cost about $70,000.[5]
Commercially available platforms were too expensive, too large, or did not meet
the performance requirements. The SRC researchers had no choice, given their
resources, but to compromise some of their performance specifications.

As of January 1989, the SRC team members had not settled on exactly
what kind of platform, given the available choices and their resources, constituted acceptable performance (yet they still hoped to have a platform by the
end of February 1989). In September 1989, the SRC members reported the
form and performance capabilities of the platform they had ordered (custom
built) but not yet received. The response characteristics of the platform were
sufficient to meet research needs at the SRC, but three degrees of freedom had
been sacrificed. Instead of one platform with six degrees-of-freedom for both
robot arms, the SRC researchers had settled on two platforms, each with one
robot, that moved on a common linear track and could rotate around two axes,
for a total of three degrees of freedom. The compromise on the platform's
capabilities also compromised the missions that the SRC would be able to simulate. By accepting the loss of two linear motions and one rotation, the SRC
researchers accepted the inability to do the demonstration they had proposed.
Without a six degree-of-freedom motion platform, they could not accurately
simulate assembly, disassembly, and repair by a two-arm robotic platform floating unfettered in space.

The original proposal initiated a developmental path for the robotic testbed; the compromises that the SRC researchers made altered that developmental path. The purchase of the platform, which was not delivered until April
1990, over 21 months after the grant began, solidified the form of the robotic
testbed and turned the performance compromises into constraints to which the
SRC researchers would have to adapt their work. Any research involving the
robotic testbed had to be do-able within those boundaries to avoid having to
alter the testbed fundamentally, and alterations were not economically feasible.

Of course, the lack of a platform for the robotic testbed did not stop the
SRC researchers from doing work. Development on other aspects of the
robotic testbed continued while the researchers were awaiting delivery of the

platform. Two other major subsystems under development, visions systems and gripper design, did remain fairly independent of the platform capabilities. Vision systems definitely influenced the capabilities of the testbed, but the vision system was also somewhat more adaptable; cameras could be moved without disturbing too much else. However, the grippers were similar to the platform; their form and capabilities would determine the tasks that could be accomplished. At the beginning of 1989, the SRC researchers decided upon the fundamentals of the project they would use to demonstrate the capabilities of the robotic testbed. The assembly task agreed upon was the assembly of a truss out of struts and nodes. This task mimicked the proposed construction method for the proposed Space Station Freedom, a task in which the sponsoring organization had a vested interest.[6] Therefore, the grippers were designed and built with this assembly task in mind (as were many other subsystems). Because the grippers were designed for a specific purpose, they were not very efficient for grasping anything other than small diameter cylinders, and there were problems when grasping those as well.[7] As with the platform choice, the design of the grippers became a constraint, albeit one much easier to alter than the platform capabilities, that affected the choice of tasks, especially once the grippers were built.

By the end of 1990, the SRC researchers had acquired all of the major hardware components for their robotic testbed, although they had neither integrated them all into a single system nor developed the software required to control each subsystem. Nevertheless, the hardware had been chosen based on research interests at the SRC, missions of interest to the sponsors, and developmental boundaries that became apparent as the project proceeded. And when the various pieces of the robotic testbed were delivered and attached to each other, the instrumentality itself constrained future work as well, a situation which became apparent in 1991.

## Changing Conditions of Development

For the first two and a half years of the SRC grant, there is virtually no record of any significant feedback from sponsor personnel regarding the robotic testbed. Only the SRC Advisory Board, which did not officially represent the sponsor, seems to have met. The Advisory Board met twice, first in September 1989, before the platform had been delivered to the SRC, and again at the end of November 1990. The Advisory Board returned a report to the SRC at the very beginning of 1991. The Advisory Board members were generally impressed by the robotic testbed, although they had hoped that its development might have proceeded further by that time. However, they also offered this prophetic advice in their report: "We believe that it will be desirable to have significant [sponsor] input to [the development of the robotic testbed] in the event of major structural changes to the design of the space station, to insure that the research results remain applicable."

This advice must have been somewhat surprising to the SRC team. Until this time they appeared to have been proceeding under the assumption that they were expected to do what they had proposed, making compromises when necessary (such as in the choice of a platform). In fact, the four status reports that the SRC researchers had put together as of the beginning of 1991 retained the same structure as the initial proposal: seven sections on the various theoretical topics being studied at the SRC, followed by an eighth regarding the status of the robotic testbed. Furthermore, it does not appear that there were any plans at the SRC to change things in the near future. According to the SRC report on Center activities during 1990, the earliest completion date for the robotic testbed was five to seven years away.

As much as some of the SRC researchers might have preferred, the SRC did not exist independently of other organizations. In the spring of 1991, NASA announced that the plans for constructing the proposed space station had changed. For the researchers at the SRC, the change in the space station construction plans was significant, for suddenly the mission that justified the relevance of the robotic testbed to their sponsor had been altered, just as it had potentially achieved its final physical form. Moreover, the robotic testbed development was becoming central to almost all work at the SRC, especially since two other projects, a flexible beam and a mobile robot, had been criticized by the Advisory Board and summarily canceled. To make matters worse, portions of the testbed, specifically the grippers, had been designed for the task of truss construction, as had some of the robotic testbed's control and planning algorithms, and researchers had spent valuable resources developing struts and nodes that the robotic testbed could handle. That the robotic testbed could not accurately simulate construction in space due to the limited degrees-of-freedom of the platform, and it was not really well suited to simulating planetary construction because its motion was fairly limited provided additional constraints. The SRC researchers found themselves in a difficult position: their grant was due to end in eighteen months, and their main project no longer appeared to be relevant to their sponsor.[8]

## Rejustifying the Robotic Testbed

With the robotic testbed project in jeopardy, the SRC researchers needed to find a way to fit into their sponsor's plans, but without making major changes to it, which they had neither the desire nor the resources to do. The initial solution for the SRC members was to attempt to convince their sponsors that the robotic testbed could be used for relevant space robotics research. The next

SRC status report was released in mid-September 1991, and the emphasis was on demonstrating the value of the robotic testbed for space research and the progress that had been made on its development. Perusal of the early part of the report would lead a reader to believe that virtually no other work was underway at the SRC; the theoretical topics, so prominent in every previous report, were not mentioned until a "Future Work" section near the end.

The prominent placement and detailed discussion of the robotic testbed were notable as an attempt to justify the testbed, but the arguments were not convincing. The purpose given for the testbed was "to provide a facility for research on space-based robotic applications," but no convincing evidence followed this claim. The development of the robotic testbed had become so closely intertwined with the strut and node truss assembly demonstration that the SRC members were hard pressed to describe other space-oriented uses for it. It was possible to do some research on compensating for disturbances of the base, but even that could not be truly space-like due to the limited degrees-of-freedom of the platform. The only visible change to the robotic testbed which the SRC report mentioned was a plan to make alterations to the control architecture so that it could accept external commands from a human operator at a remote site. In some respects this was not a major change, considering that the human would merely replace the top level of the intelligent control hierarchy, which had not been completed yet. The basic attitude at the SRC seemed to be that they were doing what they had proposed (and the sponsor had agreed to), and they would continue to do so as long as the development of the robotic testbed continued to progress, although they were willing to alter the facade.

The SRC status report from September 1991 began a short negotiation between the center and sponsor personnel, which included an evaluation report of the SRC by a Technical Review Committee (TRC), dated mid-December 1991, based on a late November site visit; a letter from the Associate Director of the SRC to the Chair of the TRC (although the chair had changed since the TRC report) sent in June 1992, in response to the TRC report; a proposal for grant renewal prepared by the SRC members, describing plans for the next five years of their research program, submitted at the end of June 1992 and a Peer Review Team (PRT) evaluation report of the SRC proposal, dated July 14, 1992.

One thing became clear: the SRC and sponsor personnel could not agree upon what constituted acceptable space robotics research and how the robotic testbed should be used. To the TRC members, the robotic testbed was "generic" and not focused on space issues. Moreover, the truss construction demonstration project was only part of an assembly task from their point of view, and no longer a relevant one at that. In response, the SRC's associate director pointed to the flexibility of the robotic testbed, and the capabilities built into it that allowed the SRC researchers to test numerous and varied theoretical topics. Regarding the truss assembly demonstration project, the associate director's only mention of it in his response to the TRC claims was that the SRC researchers would work on making conditions in the laboratory more "space-like." It was, however, in the best interest of the robotic testbed development and the research at the SRC (at least in the short term) to continue to receive funding, so the SRC members made a more concerted effort to find a research program that would fit into their sponsor's goals, and they outlined their plans in the new proposal.

For the SRC researchers, abandoning the robotic testbed, as recommended by the TRC, meant taking major steps backwards on the smaller research projects that constituted the work on the robotic testbed, if not abandoning them outright. Moreover, beginning a new, more applied project as the TRC had requested would have been a distinct challenge for the SRC researchers unless they could have acquired more human and financial resources, especially considering how all of their work had come to be focused on the robotic testbed. The solution the SRC researchers presented in their grant renewal proposal was to make changes to the robotic testbed that would improve the project, potentially speed up its completion, and once again give them a demonstration project that sponsor personnel considered relevant, but without making fundamental changes to the testbed structure, including the controlling operating system. Their tack was an example of reverse adaptation in action.

The SRC researchers proposed to replace the robot arms, which were old and designed for manufacturing purposes, with arms that had more in common with the types of robots used in space. They also proposed replacing the grippers, which would allow the manipulation of more than cylindrical objects. The proposal mentioned looking at planetary construction (which made sense, since they could not fully simulate space-based construction) and making the laboratory more planet-like by varying the lighting conditions and making the construction site uneven and cluttered.

Unaltered by the SRC renewal proposal were the platform, the operating system that had been developed for it, and many of the subsystems. In addition, the SRC researchers planned to continue with their efforts to have the robotic testbed construct a tetrahedron out of struts and nodes, as an example of planetary construction. They also proposed to continue most of the significant theoretical projects, except for the ones that had been led by faculty who had left the center. The SRC researchers did plan to make more drastic changes to the robotic testbed as well—in about three more years. Once again, too much of the applied work at the SRC had been focused on strut and node construction of a tetrahedron for the researchers to switch projects rapidly, and any new grippers they acquired would not be available, much less usable, for a while.

The only really significant change to the robotic testbed development that the SRC members proposed was to abandon striving for a fully autonomous system, and instead to begin developing strategies for telesupervision.[9] While this alteration definitely represented a compromise of the initial intention of the robotic testbed development to achieve autonomous, intelligent control, it also represented an operational threshold that could likely be achieved sooner, and it fit better within the normal operating procedures of the sponsoring organization, which did not often use fully autonomous systems. Overall, the proposal that the SRC researchers prepared for their sponsor demonstrated that their willingness to change the development plans of the robotic testbed were limited to alterations in accessories; they offered no fundamental changes which would prevent them from eventually completing the individual research projects they were involved in (and had originally proposed).

In negotiations proposals are frequently followed by counter-proposals; a six-member Peer Review Team (PRT) prepared and presented the sponsor's counter-proposal to the SRC. The PRT did not give the new SRC proposal the same rubber stamp of approval that the original had received, but they did not haphazardly discard it either. In a nut shell, the PRT wanted the SRC researchers to integrate their own activities into ongoing research at other space-oriented research centers. For the robotic testbed development, this meant importing systems that outside researchers had already developed over to the SRC, rather than having the SRC researchers develop new ones. This would insure compatibility with the sponsor's systems, but came with no guarantee of allowing the SRC researchers to continue their own research on the robotic testbed. All things considered, the PRT decided to recommend that the SRC grant be renewed, but conditionally.

The conditions upon which the grant depended were changes in the SRC program. However, before the SRC researchers had a chance to respond, another contingency arose. The SEC program was going to receive a smaller budget in fiscal year 1993 than it had in previous years, and the decision not to renew the SRC grant was made in late August 1992. The exact reason that funding was not renewed is still a matter of debate, as I found through interviews.[10] The SRC associate director believed it was because they did not have a political ally to fight for them. One review committee member believed that an administrator at the sponsoring organization needed to demonstrate his authority and used the SRC as an example. The official story from the sponsor is that due to budget constraints they either had to reduce funding to all Centers in the program or eliminate one. According to the SEC program director, the SRC stood out as the obvious choice for elimination based on its review. Realistically, it is likely that there is merit to all of these points of view, but regardless of the exact cause, the result for the SRC was the loss of their only significant source of funding and the eventual dissolution of the center. One of the many unfortunate effects of the loss of funding for the SRC program was that the negotiations between the two organizations regarding the research at the SRC and the form, function, and uses of the robotic testbed suddenly halted. The robotic testbed, like the SRC, did not suddenly cease to exist, nor were plans for its continued development canceled; however, the SRC researchers were unable to find a significant source of funding, and the Center closed for good at the end of 1993.

The form and function of the robotic testbed differ from what had originally been conceived in the first SRC proposal. Developmental boundaries on the robotic testbed project spurred many of the changes between the early form of the testbed and its initial conception. The role of sponsor personnel was minimal in the early shaping of the robotic testbed, in part because they had readily accepted the initial conception. As programs and plans changed inside the sponsoring organization, however, the SRC engineers were requested to alter their program as well. By that time the robotic testbed had attained a definite form, and had itself become a constraint to which the researchers at the SRC adapted their work. What followed were the beginnings of a renegotiation of the future form and function of the robotic testbed and research at the SRC. These renegotiations were suddenly terminated before any agreements had been reached.

## THE ROLE OF THE ROBOTIC TESTBED AS A NON-OPERATIONAL INSTRUMENTALITY IN THE SRC

I have subdivided my examination of the role of the robotic testbed as a non-operational instrumentality at the SRC into two categories: symbolic uses and research production. Symbolically, the SRC researchers used the robotic testbed to attempt to obtain more resources and to legitimate other projects. Meanwhile, the SRC researchers organized the developmental plan of the robotic testbed to allow themselves to produce output from their efforts. While some of the symbolic roles that existed for the robotic testbed affected research only indirectly, others overlapped with the phenomena that I have labeled "research production." For example, some of the smaller research projects at the SRC were undertaken only because they were portions of the robotic testbed development. Other research projects at the SRC were already underway when the SRC was formed and were subsequently reoriented towards the development of the robotic testbed. Thus, some of the more prominent research programs (intelligent control theory and task and assembly planning are good examples) were symbolically linked to the robotic testbed development to justify funding their continued development under the grant when it began, while other projects in the same programs were also directly related to getting the robotic testbed to function as desired. I will begin by looking at how the SRC researchers used the robotic testbed to represent the SRC to the rest of the world, how they used it to represent the relationship between the SRC and the sponsoring organization, and how they used it to organize and justify portions of the SRC research program, all of which were intertwined, before I explain how researchers used the robotic testbed to produce research output.

## Symbolic Uses: Obtaining Resources, Attachment, Legitimacy

Symbolically the SRC researchers used the robotic testbed in multiple and varied ways to obtain resources and legitimate their work. The robotic testbed was used as an emblem to represent the research at the SRC to external actors. Its development was used to justify many of the research projects at the SRC to sponsor personnel, and the development of the robotic testbed and the demonstration project represented the link between the two organizations. What makes the term *symbolic* appropriate is that the robotic testbed was used to represent potential in many cases, to create an impression of what could happen rather than what was happening. One reason that the robotic testbed could be used to represent the SRC was the scale of the project. Almost every project underway at the SRC was related to the robotic testbed development in one way or another, so the testbed could summarily represent research at the SRC. Individual research project results from the SRC were shared with the engineering research community through journal papers and conference presentations. However, papers and presentations generally represent the work of a small number of people (two seemed to be most common in the circles the SRC researchers moved in) rather than the entire center.

A better indication of how the center members tried to represent the center as a whole can be found by examining public relations materials. The type of glossy brochure I speak of might seem like a strange place to look for what goes on inside a research center, since there is rarely anything but the most superficial description of the research that takes place therein. Yet such descriptions are the center's face shown to the outside world and serve to attract resources such as new members and new sponsors. For example, it was this very type of brochure that was sent to prospective graduate students to attract them to the center. It could also be shown to potential sponsors to introduce them to the rudiments of the laboratory and its research. In such brochures, it is possible to create an idealized image of the laboratory that leaves the actual status of research projects unclear. The SRC was featured in two brochures (both produced in 1991), one of its own, and one produced by the sponsors. In both cases, the robotic testbed was prominently featured, along with the truss construction project. What made the brochures I refer to effective for the SRC was the use of specific photographs combined with generic descriptions of the laboratory that were related to, but did not explicitly explain the photographs. Thus the robotic testbed and truss construction project were shown posed as if they were working, which thereby implied more had been accomplished than actually had been at the time.

One notable difference between the two brochures was that in the one produced by the SRC the robotic testbed was emphasized over the truss assembly demonstration project, while the opposite was true in the brochure produced by the sponsors. It seems that the SRC researchers felt that the potential of the robotic testbed to examine various areas was more useful for attracting new members and funds, while the final product was of more interest to the sponsor and anyone to whom they might have explained the SEC program. These were basically the two views that the Associate Director of the SRC and the TRC Chair, respectively, expressed after the 1991 review of the SRC. The robotic testbed and the truss assembly demonstration project were barely separable; the combination of them represented the link between the two organizations.

Interaction between the SRC and the sponsor included exchanges of communication, money, and even personnel at times (only from the SRC to the sponsor, however), but the robotic testbed and the truss assembly demonstration project symbolized the link for most of the people involved. This was part of the role of the robotic testbed as a boundary object. For the various sponsor personnel who interacted with the SRC, the robotic testbed work at the SRC represented a tangible product that could be transferred to the sponsor and potentially used on a space mission at some time in the future. For the SRC researchers, the robotic testbed was what made their research space-oriented. Most of the smaller project work at the SRC was generic enough that it could have been adapted to some other application, such as manufacturing (a point of contention between reviewers and the SRC researchers, for while sponsor personnel wanted the SRC researchers to develop industrial ties, they wanted specific, space robotics ties). By applying research projects at the SRC to the robotic testbed development and space-truss assembly, the SRC researchers could claim that they were making unique contributions to space missions.[11]

Sponsor personnel seem only to have been convinced that the work at the SRC was sufficiently space-oriented when truss construction in space was of direct relevance to their organization's immediate plans. When the space station construction plans were altered, the SRC researchers continued to pursue truss construction as a demonstration (changing the demonstration project was not a simple task, for reasons already mentioned). Sponsor personnel, therefore, began to question the value of the robotic testbed and the theoretical projects for space research. These engineers apparently did not consider the algorithms and software for the intelligent control of a robot to be a tangible product; they desired something from the SRC researchers that could be sent up on a space mission in the near future. This view was prominent, despite the fact that by accepting the original SRC proposal, sponsor personnel had accepted a plan under which algorithms and software were the transferable products. Even if the SRC researchers had completed their development of the robotic testbed during the first grant period, the testbed itself certainly was not going to be transferred to a different research center, much less sent into space on a shuttle flight, which leads me to question just how carefully sponsor personnel explored the expected returns of the original proposals and ponder what other factors might have been involved in the selection of SEC Center sites.

The inability of the SRC researchers to demonstrate a task that sponsor members considered relevant was one of the major factors in the mediocre review that the SRC received when trying to renew their grant. As long as the robotic testbed was associated with an immediately applicable project, however, it was acceptable, and it represented a link between research at the SRC and NASA missions. This was true in spite of the fact that the capabilities of the testbed were, at their best, too limited to actually simulate construction in orbit accurately. Thus, when the robotic testbed was still being well received, it was used to justify other, smaller research projects at the SRC.

The immediate question is: justify in what sense, and to whom? In this case I use the term *justify* to signify the ability of the researchers at the SRC to legitimate their work and reassure their sponsors that various research projects were relevant to portions of the robotic testbed development, and therefore should be supported. The robotic testbed was a collection of smaller research projects that were to be orchestrated and integrated into a single large and complex instrumentality. Some of the projects were dictated by the form and function of the robotic testbed: to manipulate struts with robot arms, some form of grippers was necessary as well as some way to control those grippers. Other projects existed prior to the beginning of the SRC grant and the decision to concentrate on space-truss assembly. From one point of view, these projects were adapted to the robotic testbed development work. From another point of view, the robotic testbed development was used as a means to fund the continuation of these projects.

Both views are correct to some degree. Various research projects from the RAL were adapted to fit into the robotic testbed development project. They were not, however, the only available means to accomplish the tasks at hand. Even within the constraints of the SRC intelligent control structure (itself an ongoing research topic) and the desired form and capabilities of the robotic test-bed, there were multiple choices available for the form and function of various subsystems that were to make up the intelligently controlled robotic testbed. Many of the subsystem forms were, it seems, chosen because they were familiar and grounded in ongoing research, not because they were thought best for accomplishing the project in the most efficient or space-like manner, but because the SRC researchers had additional commitments beyond those to their sponsor.

The developmental plan the SRC researchers adopted for the robotic testbed reflected their varied goals and organizational and institutional commitments. As part of this compromise, the robotic testbed development was used as a means to allow the continued funding of research projects that were already underway when the SRC grant began. Three research projects that serve as good examples are adaptive control, petri nets, and assembly planning. All three were research topics being studied in the RAL that were adapted to the SRC mission and continued. Along with being potentially useful to the robotic testbed development, these particular projects also got three senior faculty members involved in the SRC, including a department chair. Adaptive control was one of many classes of algorithms that could have been explored as means to design controllers for the robot arms on the robotic testbed (several other controller design techniques were also explored and used during the testbed development, and adaptive control was not used for the primary controller on the robotic testbed). Petri nets and assembly planning were used in conjunction with each other on the robotic testbed as methods to plan, represent, and evaluate the truss assembly task autonomously. Originally, however, they were initially the projects of different researchers, and neither was uniquely qualified for the task of planning space-truss assembly.

One way to observe how the various projects were adapted to the robotic testbed development (without getting into technical details) is to compare the titles of RAL reports and the early SRC reports with the later SRC reports. Once funding began for the SRC, the overwhelming majority of reports issued by researchers who were members of both the SRC and the RAL (which became something of a paper entity as virtually all work and resources were attributed and assigned to the SRC) were issued as SRC reports rather than RAL reports (109 versus 11 from July 1988 to December 1991) regardless of their apparent applicability to space issues.

A brief examination of adaptive control publications provides a good example of the transition from RAL to SRC research. Adaptive control was a method being studied in the RAL, where the applications were oriented toward biomedical technologies. An RAL report and the early SRC report, respectively titled *Multiple Model Adaptive Control Method for Blood Pressure Control Systems* (1984) and *A Hybrid Adaptive Approach for Drug Delivery System* (1989), demonstrate the direction of the adaptive control research when the SRC grant began. Later, adaptive control appeared once again, now oriented toward robotics, in such reports as *Model Reference Adaptive Control of Flexible Robots in the Presence of Sudden Load Changes* (1990) and *Model Reference Adaptive Control of Robots* (1991) as well as others. One faculty member is co-author on all but one report, which was a thesis by one of his graduate students. The orientation of applications changed, but the underlying research program continued unabated. The situations with petri nets and assembly planning were virtually identical.

Linking research programs that were already underway to the robotic testbed development allowed researchers to continue work on them; the robotic testbed project provided funding and a convenient case study to examine. This is not to say that the work was detrimental to the development of the robotic testbed; these were not the only ways to approach development issues either. The strategy also had its drawbacks; it proved to be difficult to convince manufacturing firms of the usefulness of algorithms that concentrated on space-truss assembly as an application. Nonetheless, the robotic testbed development was used symbolically for researchers to legitimate their work. Thus, the symbolic uses of the robotic testbed were closely related to research production.

## RESEARCH OUTPUT: PRODUCING PAPERS AND GRADUATING STUDENTS

Before discussing how the SRC researchers used their nonoperational instrumentality to produce research output, allow me to briefly explain what could constitute research output from an academic laboratory such as the SRC. I break research output into three interconnected categories: instrumentalities and artifacts, researchers (mostly graduate students), and publications. Instrumentalities and artifacts are things—technologies and techniques—that can be exported from a research laboratory to other settings (which is not to say that they will remain unchanged or serve the same purpose in a different setting). Technologies and techniques are very tangible; sponsors or customers can export them from a research laboratory and use them on their own. This also means that they can be patented or copyrighted and licensed for others to use, usually for a price, which can bring an influx of money to a laboratory as well. People, especially students who are trained to do research, are important outputs for an academic research center. As long as a research center exists as part of an institution of higher learning, training students will be one of the most important missions of the members of the laboratory. Furthermore, students take techniques, ideas, expertise, and assumptions from a research laboratory when they leave it, a fact not lost on the sponsors, who were hoping to disseminate space-based research interests.

Publications remain somewhat distinct from instrumentalities, artifacts, and people in this sense. Although they also reflect the context of their creation, they are decontextualized (or maybe recontextualized is more accurate) by their authors before being sent out of the research laboratory (for examples see Knorr-Cetina 1981; Zenzen and Restivo 1982; Collins 1990). Publications are generally reports of findings from a research laboratory (derivations, results, theories) that have been universalized and sterilized for presentation to the larger research community. While many technical publications can be utterly incomprehensible to readers who are not familiar with the general research area, such publications are frequently written as if the location and circumstances of their development were inconsequential. Nonetheless, publications are a significant form of communication between researchers in a field, even if there is rarely enough information contained in one to duplicate an experiment or result.

Publications are still strongly related to other forms of academic research laboratory output. The completion or use of an instrumentality or artifact can be reported, as can the work graduate students must complete to finish theses and dissertations. For their part, publications from SECs were of lesser value to the sponsor's engineers and administrators than other forms of research output from the various centers. Instrumentalities, artifacts, and researchers were more tangible, and therefore immediately useful. For the researchers at the SRC, who were part of a different community than the sponsor personnel, all forms of output were important to the point that they could not afford to have the extended development of the robotic testbed halt other forms of production.

What the SRC researchers did was use their developmental efforts on the robotic testbed to produce publications and help students complete their graduate work. This points to the difference between using an instrumentality for research production and using the development of an instrumentality to produce research. For the SRC researchers, the robotic testbed was to be a testbed for their theories of intelligent control: how intelligent control might work and under what conditions, at least for autonomous construction (of space-trusses). However, the work on an intelligent controller for the robotic testbed was not completed during the period when the SRC was sponsored by a significant grant. The SRC intelligent control structure was hierarchical, and most of the work was accomplished for the lower portions, but the upper end remained incomplete or nonexistent as part of the robotic testbed. Thus demonstrating and examining intelligently controlled autonomous construction never became an option for the members of the SRC.

The modus operandi of turning developmental efforts into research output was not complicated. Once a task was identified, such as the design of grippers or a controller for the platform for the robotic testbed, one or more researchers were assigned to the project. Many of the projects ended up at the level of Master of Science theses or Master of Engineering projects, and therefore could be assigned to a single graduate student (and faculty advisor). Thus the work of a student (for which the student likely received some form of funding) served to advance the development of the robotic testbed, while at the same time providing the student with the thesis or project required to graduate. Moreover, a completed thesis could also be published as an SRC report, frequently as a conference presentation, and occasionally in a peer-reviewed journal (usually only from Ph.D.-level work however). Furthermore, if a developmental task was too large or complex for one or two people, or just would take too long, it could be subdivided among several students, either in parallel or in series. For example, in the course of getting grippers for the robotic testbed designed, operating, and integrated into the robotic testbed, two students fulfilled requirements for graduation, and one conference paper and three reports were produced. Similar examples can be found with almost every subject, including control of the testbed, vision systems, assembly planning, and theoretical work on intelligent control.

There is, not surprisingly, a great deal of overlap between theses, the SRC reports, and conference and journal papers. In some cases, the work in one dissertation ended up as several SRC reports, several conference papers, and one or two peer-reviewed journal articles. This is not an unusual practice, but it does help inflate the publication numbers. Approximately two-thirds of all the SRC reports were presented at conferences, published in peer-reviewed journals, both presented at a conference and later published in a journal, or a variation of the work in the report was disseminated by one of the aforementioned means. It is important to note, however, that the applied work on the robotic testbed development was much less likely to lead to a publication, other than an SRC report, than more theoretical work, and only rarely did an applied task generate more than a conference presentation. This was in part a function of the nature of the work that was being reported: more applied tasks were given to junior students, while more advanced students undertook more theoretical missions. But the form of the academic community was also important, since the results of applied work were generally considered worthy of publication only in conference proceedings, and not in peer-reviewed journals. For instance, the work of three students led to one conference presentation in the case of the grippers for the robotic testbed, while the work of one doctoral student on the reliability of intelligent machines (anything but an applied topic) led to two conference presentations, one journal article, and one book chapter.

The bottom line is that using the development of the robotic testbed was not the most efficient way for the SRC members to produce publications. Moreover, subdividing the robotic testbed development into tasks that single students could tackle was arguably not the most efficient way to get the testbed to its operational threshold; the best way to train researchers is not necessarily the best way to use labor. However, the SRC researchers established a balance that reflected their multiple commitments. It allowed them to produce publications, have students graduate in a reasonable period of time, and continue developmental work on the robotic testbed. This was done rather than suspending theoretical work that was not associated with the robotic testbed development or forcing students to do developmental work for their stipends independent of their theses. Thus, no one form of research production was sacrificed for the sake of the others. This reflects the institutional setting in which the SRC researchers resided. Along with satisfying the desires of their sponsor, the SRC researchers also had to meet university goals (both school-wide and inside the School of Engineering), academic norms, and student expectations.

Using the development of the robotic testbed to produce publications and keep students efficiently moving through their graduate programs was one of the ways that the SRC researchers used the robotic testbed. The SRC researchers also used the robotic testbed as a symbol to represent their relationship with their sponsor. The nonoperational robotic testbed was used to justify the continuation of research projects that predated the SRC and to represent the Center to those not involved, whereby new members and sponsors were to be attracted. In the end, the robotic testbed was central to most research and nonresearch activities that took place inside the SRC, regardless of the fact that the researchers still had a great deal of development to accomplish before the robotic testbed could become operational.

## Reflections on the SRC Experience

It is unclear how the form of the robotic testbed might have changed had the SRC-sponsor relationship continued beyond the fall of 1992. Moreover, it is also uncertain how many more months or years of development would have been required to bring the robotic testbed under autonomous, intelligent control, or what the SRC researchers would have been able to do with the testbed, had such an event ever occurred. The development of the robotic testbed was a risky venture for the SRC researchers, although they may not have realized so at the time they proposed it. The SRC researchers may have had real confidence in their ability to put their intelligent control theories into practice and demonstrate a myriad of uses for them, but confidence is no guarantee of success.

One could surmise that the SRC researchers simply did poor research, but by the standards of the academic community they were very productive. The renewal proposal cited 326 publications by SRC personnel, including six books and eighty-three articles in refereed journals, not to mention more than one hundred technical reports. In addition, fifty-four students received advanced degrees (thirteen received doctorates) after doing research in the center. From the perspective of satisfying their commitments to the university, the academic community, and their students, the SRC was an extremely productive research center. Moreover, the SRC researchers tried to do exactly what they had originally proposed, and when they were asked to make changes, they did what they could without forsaking their other responsibilities. Their decision to equate strut and node truss construction (originally of interest to the sponsor) with space-based construction affected other decisions, such as the type of grippers that were designed for the robot arms and many of the algorithms to control the robotic testbed. None of these features was unalterable, but given the financial and human resources of the SRC and the effect of changes to the robotic testbed on smaller research programs, significant changes to the testbed were neither feasible nor desirable for the SRC researchers.

The case of the robotic testbed development project at the SRC demonstrated some of the forms that the negotiation of the development of an instrumentality can take. The case showed how successful some attempts to define a research project can be (or appear to be), as with the sponsor's researchers' ready acceptance of the original SRC proposal. But it also showed how difficult it can be for researchers to try to find a way to define and orient their work so that an agreement can be forged between researchers and sponsors. Moreover, the case was an example of how various developmental boundaries— technical, economic, physical—can affect the form and do-ability of the development of an instrumentality, and how early compromises can become embodied into the form of the instrumentality and become hindrances to future negotiations between various personnel.

The robotic testbed was an intra- and extra-organizational boundary object, and the SRC researchers used it in multiple ways. They used it as a symbol to represent the work at the center to outsiders and to represent their link to their sponsor and space research. They also used it to justify the use of the grant for funding ongoing research programs, while encapsulating portions of the developmental efforts to increase their number of publications and allow students to graduate in reasonable periods of time. For the SRC researchers, the operational status of the robotic testbed was somewhat inconsequential when it came to representing the Center. Although the representation of the robotic testbed would probably have been different had it passed its operational threshold, the SRC researchers used images of it to represent where they were going with their research, rather than where they actually were. Thus, the robotic testbed was used by the researchers at the SRC to attempt to attract new members (i.e., graduate students) and new sponsors. The center was much more successful at obtaining new researchers than new sponsors, and the link to space missions more than likely had something to do with that situation.[12]

The robotic testbed, combined with the strut assembly demonstration project, represented the link between the SRC and its sponsor. The robotic testbed and demonstration project brought all of the research at the SRC together and focused it on space-truss assembly. As long as the space-truss assembly project was relevant to the sponsor's missions, researchers at both organizations were able to agree that the research at the SRC was on space robotics. It became clear by late 1991 that researchers from the two organizations did not see eye to eye on the matter of what constituted space robotics research without a relevant demonstration project to tie everything together. Once the sponsor declared that space-truss assembly was no longer of interest to them, the robotic testbed was not "space-like" enough; meanwhile, it still represented space robotics research to the members of the SRC. On a philosophical level, researchers in both organizations had valid arguments to back up their claims regarding the space relevancy of the SRC robotic testbed. Politically, however, the sponsors had much more clout than the SRC researchers when it came to differentiating between space-oriented and other types of research. Thus, how space-oriented the robotic testbed actually was was a function of the conception of space issues by researchers at different organizations, who negotiated with different levels of authority and credibility.

Understanding the relationship between the SRC and the sponsoring organization and the form and role of the robotic testbed that was the crux of this relationship requires understanding that the SRC was based upon a continuation of pre-existing relationships, and these relationships provided much of the basis for the form and function of the robotic testbed and the organization of research inside the center. The SRC case demonstrates that the very development of an instrumentality altered the negotiations regarding its own completion. The robotic testbed was a nonoperational instrumentality, but as a focal point for organizational relationships and a physical artifact with limited capabilities, it constrained the options of the SRC researchers. This is an expansion of Price's concept of instrumentality: the implications of trying to obtain one are much more significant than the research advantage that may or may not be gained by its acquisition. This case also further demonstrates the roles that both contingency and background, including organization and institutional affiliations, can play in developing and maintaining a successful research program.

One final observation based on the SRC experiences is the difficulty and potential drawbacks under current academic research funding conditions in the United States of attempting a project that requires a long development period. The SRC researchers began work on an instrumentality they knew could take a very long time to develop (their own estimate at the beginning of 1991 was five to seven more years) at a time when sponsor personnel, who eventually desired something tangible and quickly due to the uncertainty of their own future, were planning on a faster return on their investment. Time to completion was only one of many differences between the desires of the sponsors and the SRC researchers in 1991. It does raise the question, also raised by Price (1984), Remington (1988), and others, of whether the current funding policies for academic scientific and engineering research are sufficient to allow long-term research programs, especially the development of instrumentalities, to proceed smoothly to completion. When researchers place themselves (or are placed) in academic research centers, they obtain resources, legitimate their work, and attempt to retain some autonomy over it in a space that is shaped by interactions with multiple institutions and organizations that could very well have incompatible goals. There is still room for a greater examination of what the implications of such a setting are for training students, choosing research topics, and the professional development of faculty members, especially regarding the socialization of new generations of technological researchers and the values that are imparted though such interactions.

## NOTES

This chapter is a selection from *Constructing Space: The Shaping and Uses of the CIRSEE Two-Arm Robotic Testbed* (1993), unpublished M.S. Thesis, Department of Science and Technology Studies, Rensselaer, Troy, NY.

1. I accomplished the bulk of this study through the examination of documents produced by SRC researchers and sponsor personnel who were involved with overseeing the grant. I concentrated on charting the development of the two-arm robotic testbed and examining how it was represented in various types of documents (semiannual and annual reports, theses, technical reports, correspondence, technical review committee reports, public relations brochures, and presentation slides). Specifically, I examined the roles (product, simulation platform, example, justifying factor) in which researchers used the testbed in various documents, how its roles varied over time, and how sponsor personnel reacted to its various forms.

I supplemented my examination of documents in two manners. First, I conducted a limited number of interviews with the final SRC director, three members of the sponsoring organization who were part of the team responsible for overseeing the SRC grant, and several graduate students. Second, I spent two and a half years as a graduate student in the SRC. Although I never worked directly on the two-arm robotic testbed project, I was an indigenous researcher. I was a participant-observer on another, much smaller (and shorter-lived) project, and a local observer of the robotic testbed development. Being physically present allowed me to witness events as they transpired and to become familiar with the setting and the researchers. Never being involved in the development of the two-arm robotic testbed, however, afforded me a detached standpoint with which I could view activities in the SRC, one that researchers involved in the robotic testbed's development did not have.

2. One thing which may strike readers who are familiar with David Noble's work is the uncanny similarity between the SRC experience and project Whirlwind at

MIT shortly after World War II. In both cases the academic researchers attempted to accomplish less specifically applied research under the guise of producing a specific technology for the project's sponsors, and in the end, support was withdrawn by the sponsors before the academic researchers were able to satisfactorily complete their project (Noble 1986).

3. Another way to view this situation is that the project had "momentum" (Hughes 1989).

4. The center of mass of a system with two robot arms on a single platform that is floating in space (i.e., neither anchored to another object nor under the influence of any propulsion or attitude compensation systems) will remain at a constant location (or on a constant path if it is not initially at rest). Under these conditions, the Newtonian notion that "for each action there is an equal and opposite reaction" explains the system's behavior. If one or both of the robot arms moves (unless they move in a properly oriented symmetric manner), the entire platform will move reactively in some manner so that the center of mass remains at a constant location.

5. There was also a maximum load that the laboratory floor could withstand.

6. Eventually the demonstration project, as it was commonly known, was broken into four levels: (1) insertion of a third strut into a triangle using one arm, (2) the construction of a triangle using one arm, (3) the construction of a triangle using two arms, (4) the construction of a tetrahedron using two arms. However, this plan was not described in detail (in print) until mid-September 1991, five months after the sponsor announced that the space station would not be constructed in such a manner.

7. The SRC researcher's decision to build specialized grippers was not out of line with normal robotics research. True general purpose robotic grippers that can efficiently manipulate a wide variety of object shapes are rare.

8. The SRC never managed to generate any significant support from industry, and support from the university was minimal relative to the overall operating budget.

9. Teleoperation of a robot means a human at a remote site is controlling all of its motions. One example of this is the arm on the space shuttle, which is controlled from inside the cabin. Telesupervision means a human at a remote site issues high-level commands from which the robot then makes plans and executes the mission without further intervention. One example might be issuing an order to a robot to move to a point and construct a tetrahedron, at which time the robot would plan how to move to the point and do the construction without further commands from the remote operator.

10. All other centers in the program did receive grant renewals.

11. Although most of the work at the SRC could have been adapted to industrial tasks as easily as space-oriented ones, the space "flavor" was prevalent enough at the SRC that the members were never able to garner any significant support from industry. In true catch-22 fashion, the SRC researchers could only interest industry in their work by moving away from space, but sponsor personnel wanted the SRC to convince industry to move into space research. Thus the center remained almost entirely dependent upon one sponsor for support.

12. Again, center members were really in an untenable position when it came to obtaining industrial support. To obtain money from the corporations that they had dealt with in the past (as RAL members before the inception of the SRC), they had to move away from the space aspects of their work, which was not acceptable to sponsor personnel. To satisfy sponsor personnel, however, the SRC researchers needed to find other organizations that were working in space robotics; there were not many corporations doing so, and they evidently were not interested in investing in the SRC. This, in turn, left the SRC researchers wholly dependent upon their sponsor for continued support.

# 10

# Systemic Influences: Some Effects of the World of Commerce on University Science

## DANIEL LEE KLEINMAN

To date, much of the existing research on university-industry relations (UIRs) in the biotechnology area in the United States has focused on the direct and immediate effects of industry involvement with university scientists on the culture of academic science. Attention has been aimed at restrictions on the flow of scientific information and patron roles in agenda setting. Drawing on ethnographic and documentary data, I explore the indirect, systemic and pervasive effects on the practices of academic scientists of undertaking research in an environment shaped by the U.S. intellectual property regime and an industry's historical domination of a field of scientific investigation.

The increasing commercial potential of university biology and the simultaneous growth of UIRs in the biological sciences over the past two decades or so in the United States has sparked considerable discussion about the impact these relationships can have on the culture of academic science. Two questions have been central to policy debates and university soul-searching: how do commercially motivated intellectual property protection considerations affect the flow of information among scholars, and to what extent does industry financial support for academic research undermine the relatively autonomous research agenda setting undertaken by university scientists?

One of the earliest observers of industry involvement in academic biology, Martin Kenney, drew on extensive interviews with the principles in "the university-industrial complex" and on a wide range of documentary sources. Kenney

225

(1988) predicted that "traditional values of openness and freedom will be replaced increasingly by obsessive secrecy and attempts to patent any and all discoveries considered commercializable" (110–111) as a result of the emergence of UIRs in the biotechnology area.

Today the work of Blumenthal et al. (1986a,b) continues to provide the only data from a large-scale national survey with which to examine the effects of industry involvement in university biology. They found that biology faculty with industry support were significantly more likely than their colleagues without such support to report that their research had resulted in trade secrets—12 percent as compared to 3 percent.

A more recent case study of university-industry relationships in the College of Agricultural and Life Sciences at the University of Wisconsin found few "concrete breaches" of the norms of science, but did find examples illustrating a threat to the spirit of academic research (Kleinman and Kloppenburg 1988). The quantitative assessment of Krimsky, Ennis, and Weissman (1991) led those scholars to conclude that concentration of UIRs in elite U.S. institutions is likely to have profound implications for the future of academic basic biology.

A host of studies have examined the second issue at the center of discussions about UIRs: the influence industrial patrons have on university research agendas. Nelkin and Nelson (1987) suggest that the new alliances between industry and universities "represent a significant increase in the influence of potential commercial opportunities on decisions about research priorities" (71).

Mackenzie, Keating, and Cambrosio (1990) argue that universities are under pressure to define research agendas that are of commercial interest. The Blumenthal study (1986b) confirmed this. The survey data from that project indicate that scientists with industry support were more likely to say that commercial considerations affected their choice of research than were those without such support. This sense of the research climate is echoed by Feller (1991); more recently, Webster (1994) devoted attention to outlining a framework to provide a means to assess the array of forces shaping university research agendas.

Historical research provides the most convincing evidence of the impact of nonindustrial patron support on university science. In his study of Cold War physics, for example, Leslie (1993) documents the U.S. military's role in defining disciplines, such as materials science, and in promoting specific research methodologies. Exploring a different field and a different patron, Kohler (1990) provides a detailed analysis of the impact of Rockefeller Foundation funding on the contours of early work in molecular biology.

To date, most research on university-industry relationships in the biotechnology area has focused on the direct and immediate effects of industry involvement with university scientists on academic research.[1] Analysts have been concerned with scientists whose investigations are financially supported by commercial entities. Investigators have scrutinized the relationship between university scientists' intellectual property protection practices and corporate support. Research results from studies of scientists' approaches to intellectual property protection have been understood to capture changes in the environment in which university science is undertaken.

By contrast, I am interested in exploring how the environment in which university science is done indirectly and pervasively affects the practices and decisions of academic scientists. I am not concerned with how a specific relationship between an academic scientist and a commercial enterprise shapes that scientist's research agenda or orientation toward intellectual property. Instead, I wish to consider how the practices and decisions of scientists are constrained by the institutional environment they inhabit.[2] In the case at hand, the specific dimensions of that environment with which I am concerned are the U.S. intellectual property regime and an industry's historical dominance of a field of scientific investigation.

In this chapter, I draw on data from a multisite, multimethod ethnography. I began my larger study by spending six months as a participant observer in a plant pathology laboratory at the University of Wisconsin-Madison. The laboratory leader gave me unrestricted access to all laboratory documents. I attended a wide range of laboratory meetings (including sessions with industry collaborators and discussions with university administrators whose job it is to coordinate UIRs) and participated in daily laboratory activities, aiding lab personnel in large-scale lab and field projects, coordinating a weekly lab meeting, and undertaking routine experimental work. Finally, I conducted semistructured interviews with virtually all lab personnel, including the lab leader, post-doctoral fellows, graduate students, and staff researchers. In what follows, I supplement this data with material drawn from an array of secondary sources.

This chapter is divided into four sections. First, I describe the laboratory in which I worked and explain why it is an ideal case for examining the impact of university-industry relations in the biological sciences. Next, I consider the flow of information and research materials from other laboratories into the Wisconsin lab in which I worked. Third, I explore the thoroughly pervasive influence of commonly held attitudes toward intellectual property protection. Finally, I examine the indirect effects of industry domination of a research field on scientific practice.

My argument can be summarized as follows: commentators on UIRs worry that industry involvement with academic scientists will lead to restrictions on the flow of information and materials central to the research process; however, my research suggests that even absent direct industry involvement in laboratory research, the norm of what Merton (1942/1973) referred to as scientific "communism" is, for a wide range of reasons, sometimes violated, and existing research suggests that it is a mistake for investigators to treat this ideal as strictly or typically adhered to.

In addition, if the data from my study capture more general trends from a policy perspective, it may be more appropriate and likely more profitable to be concerned less with the egregious and exceptional cases and more with how widely held assumptions about intellectual property, beyond the control of any specific university laboratory, can shape a laboratory's decisions about seeking intellectual property protection. Furthermore, insofar as the historical legacy of the research area of interest to the scientists in the lab I studied bears similarities to the historical foundations of other research areas of interest to university scientists, my investigation suggests that we should not allow concern with how direct industry financial support can shape university research agendas to lead us to overlook the ways in which the history of industry involvement in an area of investigation can indirectly shape research and development.

## THE SITE

From March through September 1995, I was a participant observer in the laboratory of Professor Jo Handelsman in the Department of Plant Pathology at the University of Wisconsin-Madison. The Handelsman laboratory provides an excellent site in which to examine the host of concerns that have surrounded the emergence of university-industry relations in the biological sciences. Although much of the laboratory's work is of the most basic variety, Handelsman and her lab are committed to doing research which, at least in the long term, will have practical benefits for Wisconsin agriculture. In addition, Professor Handelsman believes that basic biological research provides the fundamental basis for possible commercial applications.

Almost since Handelsman joined the faculty in the Department of Plant Pathology in 1985, the central, although not the sole, focus of research in the laboratory has been biological control—the use of biological materials to control plant disease. The primary object of the lab's work is a strain of *Bacillus cereus*, referred to as UW85, which, laboratory research indicates, contributes to the suppression of disease that affects a range of agriculturally important crops, most centrally alfalfa and soybeans (Handelsman et al. 1990; Osburn et al. 1995).

For the study of university-industry relations, the Handelsman lab is a complicated case. On the one hand, the laboratory is intimately involved with industry and the commercial world. Over the past decade, the lab has received financial support from a host of companies, including Cargill, Liphatech, and Gustafson. In addition, the world of industry norms exists in the laboratory in the sense that Handelsman is alert to the possibilities and benefits of legal intellectual property protection, and, as of July 1996, work from the lab had resulted in three U.S. patents, two European Patent Office patents, and some eighteen additional patents issued by countries around the world.

On the other hand, the Handelsman lab poses a useful though difficult case. Handelsman is absolutely clear that her first responsibility is to teach her students to be careful and creative scientists. She is scrupulous in her efforts to prevent industry contact from adversely affecting the education of her students. She does not accept contract research. Industry gifts received by her lab can have no restrictions placed upon them. On questions of intellectual property, her priority is always facilitating the dissemination of research results throughout the scientific community.

Given Handelsman's high level of self-consciousness, one would not expect to find the most obvious and egregious problems that can arise from UIRs. On the other hand, one might safely assume that modes of commercial influence uncovered are likely to be as common or more common in labs in which the leader takes a less reflective and proactive role than Professor Handelsman.

## THE NORMS OF SCIENCE: THEORY AND PRACTICE

Over a half century after he first articulated it, Merton's (1942/1973) delineation of the norms of science stands as the clearest statement of the orientation that is supposed to mark science as a distinctive social institution. Central to the normative structure of science, according to Merton, is what he refers to as scientific "communism." For Merton, the term means the "common ownership of goods" (273). According to this norm, "The substantive findings of science are a product of social collaboration and are assigned to the community. They constitute a common heritage . . . . Property rights in science are whittled down to a bare minimum by rationale of the scientific ethic" (273). According to this position, secrecy is strictly forbidden, and according to Merton, the capitalist patent system and private property in technology are incompatible with this aspect of the scientific ethos (Merton).

The logic of Merton's conceptualization has been sharply criticized, and a good deal of research challenges the empirical reality of the ethos of science. Conceptually, for example, Mulkay (1980) shows that it is not reasonable to assume that a given norm has a single literal meaning. Instead, he suggests that interpretation of norms is context-dependent, and thus violation of the literal meaning of norms is likely in practice. Empirically, Mitroff (1974), for example, in his investigation of scientists doing moon-related research, points to the counter-norm of secrecy operating as a balance to the norm of communism.

Despite the wealth of critical research on the circulation and operation of these traditional norms, the earliest commentators on UIRs in biotechnology treated these new relationships as virtually an unprecedented threat to the

norms of science. For example, in his discussion of university-industry cooperation, Giamatti (1982), former Yale University president, wrote of the tension
between the "academic imperative to seek knowledge objectively and to share
it openly and freely" and the contradictory goal in industrial science to "treat
knowledge as private property" (1279). Even those who recognize the ques-
tionable reality of these norms in the university context have commented on
the threat to scientific openness posed by the rapid expansion of the "university-industrial complex" (Kenney 1986,108–111). What is more, according to
one analyst, "it is fair to say that many scientists believe that these norms guide
their practice" (Rabinow 1996,13).

In my six months of participant observation in the Handelsman lab and a
subsequent year of monitoring the lab, I found no evidence that the laboratory's relationship with industrial collaborators or the desire of lab personnel to
"commodify" laboratory inventions led to specific instances of secrecy or
restrictions on the flow of laboratory materials to those outside the lab. In laboratory correspondence, Handelsman keeps the Wisconsin Alumni Research
Foundation (WARF), the University's patent agent, apprised of the lab's intentions concerning presentation of laboratory research at scientific meetings or
plans to submit publications, but there is no evidence that a significant restric-
tion in the flow of information resulted from this cooperative relationship. In
fact, as I noted earlier, on several occasions where related issues were raised at
meetings with graduate students, Handelsman was clear that communication
with other scientists was always the lab's priority.

Still, if the free flow of information has remained relatively unimpeded by
the Handelsman lab's interest in patenting and commercial development of lab
inventions, laboratory activities have on occasion been hampered by restrictions running in the other direction, that is, from outside the lab into it. In one
instance, a graduate student sent his bacterial strains to a company that per-
forms analysis to determine a particular type of biochemical profile of bacterial
strains. When he indicated to company staff members that he was unhappy
with the work that had been done, a senior company official forbade him from
using the results in published material.

In another instance, a Ph.D. student contacted a scientist and requested a
particular *E. coli* strain to which the scientist referred in a published paper.
When the material arrived and the student found that it was not alive, he contacted the scientist again, requesting a fresh sample. He received no response.
Finally, one lab scientist told me of a friend who could not get published strains
from a competitor, "apparently because . . . [the friend] was doing research that
could possibly undermine the competitor's conclusions."

Stories of such difficulty are common bases for frustration and humor in
the lab. In most instances, commercial motivations appear to have little to do

with information and material flow restrictions. Sometimes the explanation may be interlab competition. In other cases, as one lab worker suggested, failure to maintain biological samples properly may make it impossible for scientists to respond to requests for samples. In still other cases, it may be simple laziness.

Beyond the Handelsman lab, conflicts within research teams (Mishkin 1995, 927), nationalism (Cohen 1997, 1961), competition within the scientific community, and cost (Marshall 1997, 525) are among the explanations given by scientists and analysts for restrictions in the flow of research materials. In a recent survey, 46 percent of life scientistrespondents said they had withheld data and/or materials from other academics to protect their scientific lead, 27 percent said cost affected their decision to withhold data and/or materials, and only 24 percent referred to financial interest and/or agreement with a company as affecting their decision (Marshall 1997, 525).

What is important is that although openness may be an ideal in science, this aim is often not fully realized, and frequently the failure to achieve the goal may have little to do with direct industry influence on university science or the commercial motivations of academic scientists. In the case of the Handelsman lab, it is impossible to know for certain what prompted other labs' lax responsiveness to requests for information and materials. Nevertheless, one can certainly imagine barriers, whether they are systematic (e.g., the pressures of the "tenure clock") or ad hoc (e.g., failure to maintain research materials properly), that predate the introduction of UIRs to academic biology.

## THE IDEOLOGY OF INTELLECTUAL PROPERTY

Although it occupies a relatively limited amount of researchers' time, patenting inventions derived from lab research is a regular practice in the Handelsman lab. However, this is not a money-hungry group, and individual gain is not prominent among the reasons that lab personnel use to explain why they patent laboratory inventions.

In formal interview contexts and informal discussions, Ph.D. students, lab scientists, and Handelsman herself listed three primary reasons to pursue patents on lab technology. First, lab personnel are interested in their work having social usefulness, and Handelsman suggests that without a patent companies are not likely to be interested in commercializing lab inventions.

Second, lab workers believe that patenting may give the laboratory greater control over the development of their work into commercial products than they would have otherwise. Along these lines, one lab researcher said that if Handelsman patents an invention, then "she's got sort of ultimate freedom to do with that discovery whatever she wants." Another noted that patenting "gives the creators of something more control over how it is developed."

Finally, in times of uncertainty about the availability of federal research funding, lab researchers see patenting as a potential source of revenue to support further lab research. In this connection, one Ph.D. student noted that patenting is "good financially for the lab. . . . [T]he lab gets a certain percentage and then that can help fund other projects." Along these lines, another researcher suggested that patenting "directs proceeds . . . back in the direction of where useful things came from."

These reasons fit into commonly accepted notions about how and why the system of intellectual property protection in the United States works. The U.S. Constitution points to the granting of monopoly on intellectual property as a means to promote "practical applications." Early proponents of patenting by university scientists stressed the control over their inventions that the patent afforded (Weiner 1987, 50; Blumenthal, Epstein, and Maxwell 1986, 1622). Of course, the possibility of a cash reward, whether individual or collective, is typically considered the ultimate motivation for patenting.

What is interesting is that there is no clear empirical evidence to support several of these beliefs. Dworkin (1987) stresses the virtual absence of data bolstering the basic assertion that the patent system promotes innovation and the widespread use of invention. Buttel and Belsky (1987, 35) go so far as to assert that in the case of plant patent law, there is affirmative evidence suggesting that intellectual property is not necessary to promote innovation and commercialization (see also Kloppenburg 1988).

Early in this century some analysts asserted that patenting is unnecessary to protect the public interest (Weiner 1987, 54, 44). In addition, evidence from correspondence between Handelsman and WARF suggests that once WARF is assigned a patent, researcher control over invention development and commercialization may be limited. According to this correspondence, a company licensing a lab invention need not provide the lab with information concerning the use to which the invention will be put.

In addition, the breadth of lab patents and licensing arrangements negotiated by WARF can restrict, as it appears to have in one instance, the lab's flexibility in seeking further corporate financial support. The breadth of a patent licensed to one company may prevent Handelsman from using the possibility of a future related invention as a way to interest another company in providing support for the lab's research.

Although providing financial support for continued research in the laboratory seems a worthy goal, the history of patenting at the University of Wisconsin makes realization of this goal seem unlikely to any significant extent. Between its founding in 1925 and 1986, WARF received 2,400 disclosures from faculty. Of those fewer than 20 percent were patented. Only ninety-five patents resulted in licenses, and only seventy-six produced net income. In short, only 3 percent of WARF patents between 1925 and 1986 produced income, and only 1 percent earned more than $100,000 (Blumenthal et al. 1986, 1624).[3] Speaking more generally, Derek Bok suggests that it typically "takes one thousand reported discoveries to produce one hundred patents; it takes one hundred patents to produce ten licenses; and only one license in ten will yield more than $25,000 a year" (in *American Association of University Professors* 1983, 21a).

From a policy perspective, undoubtedly the most important consideration is the extent to which intellectual property protection contributes to innovation. Unfortunately, this is the most difficult question to answer because any demonstration of intellectual property law's contribution to technological advance would need to be largely counterfactual (cf. Boyle 1996, 205, n.11; Samuelson 1987), that is, it is difficult to do an analysis that assumes rates of technological development in the absence of intellectual property protection.

It might be argued that the reality of the situation is inconsequential, since if people believe that such a relationship exists, they will act as if it does (Samuelson 1987, 12). On the other hand, one recent analysis suggests that intellectual property law may actually hinder technological innovation. This argument centers on the role of the public domain in promoting innovation, and it is forcefully made by Boyle (1996).

Boyle (1996) contends that intellectual property law is undergirded by a romantic nineteenth-century notion of authorship: innovation is produced by autonomous individuals. This premise leads us to undervalue the role of sources and audiences in the innovation process. As Boyle puts it: "The author vision blinds us to the importance of the commons—to the importance of the raw material from which information products are constructed" (xiv).

In addition, Boyle (1996) suggests that the "author vision" leads us to "fail to consider the extent to which innovators can recover their investment by methods other than intellectual property—packaging, reputation, being first to market, trading on knowledge of the likely economic effects of the innovation and so on" (140). Finally, he warns that it does not follow from the belief that private industry will produce useful products more quickly than state-supported institutions that "the greater the level of intellectual property protection, the greater the progress" (10).

The point of this discussion is not to suggest that the Handelsman lab and labs like it are mistaken in pursuing patents on their inventions and may be doing more social harm than good in such pursuit. Indeed, if companies in a position to develop laboratory inventions believe in the efficacy of intellectual property protection, researchers may have no alternative but to seek patent protection if they wish to see their inventions commercially developed. In short, institutionalized myths (Myer and Rowan 1977), rather than rational calculation of economic efficacy, must guide the actions of academic scientists seeking legitimacy and right of participation in the commercial world. Their practices must become isomorphic (DiMaggio and Powell 1983) with those of the institutional environment in which they hope to operate.

The realities academic scientists interested in commercializing their work confront do not lead me to argue that we should scrap the existing intellectual property regime since its efficacy is at best not proven and at worst may be detrimental to technological progress. Instead, I want to suggest that focus on the impact on information flow among researchers resulting from increased university-industry contact may be diverting us from a more important consideration: the utility of our existing system of intellectual property protection. In the Handelsman lab and elsewhere, information flow problems will likely exist even in the absence of UIRs, and while university administrators are busy crafting guidelines to deal with the reality of UIRs and working to heighten the awareness of faculty to the importance of open communication, administrators, faculty, and policymakers are ignoring the need for a thoroughgoing reassessment of our system of intellectual property protection.

## BEYOND UIRS: THE PERVASIVE INFLUENCE OF INDUSTRY

Among the concerns most difficult to track and most regularly expressed by critics of university-industry relations is that strategically utilized industry funds can shape the research agendas of university scientists. Part of the difficulty in pinpointing the influence of industry in defining research programs is in understanding what researchers would have done under different circumstances.

In any case, in the era of big science, the image of a free market of ideas, in which the "objectively" best and most important problems are investigated, presented by the mythmakers of the scientific community like Michael Polanyi (1951, 1962) probably never existed. Whether their role is coercive or cooperative, across the recent history of science, patrons have played a role in defining what research is undertaken (Kohler 1990; Leslie 1993; Noble 1984).

However, a problem that has been largely overlooked in the debate about university-industry relations is the systemic influence that industry or other patrons can have through their history of dominance of a research field. In this connection, the Handelsman lab provides a very interesting case. For while Handelsman's self-reflectiveness about the influence of industry on the work of her lab can buffer the lab from the kind of effects most commonly explored in the debates about university-industry relations, the mark of industry on past research in a given area of exploration is not so easily circumvented. The findings of this research can become an "obligatory point of passage" (Latour 1987,

1988) for future academic research. That is, research in an area shaped by industry must draw on the terms, methods, measures, and findings resulting from earlier industrial investigation. The foundation established by earlier industry research may facilitate future investigation, but is also likely to be constraining.[4]

Work in the Handelsman lab, although not shaped by the purse of any agrichemical industry patron, is affected by the dominant role, since at least the Second World War, of the chemical industry in defining disease control strategies in agriculture. Commercially developed organic chemical pesticides constituted 90 percent of the U.S. market for agricultural pest protection by the mid-1950s (Kohn 1987, 162, 163) and continue to dominate the market today.

As a result of industry dominance of agricultural pest protection, what it means for UW85 to be an effective biocontrol agent is defined in terms of the chemical currently used to protect the crops on which UW85 is being tested.[5] The chemical commonly recommended is metalaxyl, a fungicide manufactured by Ciba-Geigy and approved for widespread use in 1983.

It is not direct agrichemical company funding of work in the Handelsman lab that affects laboratory research priorities. Instead, industry's historical domination of agricultural pest protection as a research field and a commercial market indirectly affects the Handelsman lab's research. Industry has shaped the character of the research area. Thus, experimental design is shaped by metalaxyl's commercial status, and research results are reported for academic audiences in terms of the comparative performance of metalaxyl and UW85.

In tests of UW85 in the field on soybeans and alfalfa by the Handelsman lab and its collaborators, the effectiveness of the *Bacillus cereus* strain in suppressing the fungi that cause damping off and root rot is measured against various formulations and applications of metalaxyl. The peer reviewed article that the lab has published on its field trials with soybeans shows that the benefits of UW85 are similar to those of metalaxyl (Osburn et al. 1995).

Significantly, past research on metalaxyl and the consequent knowledge of its target specificity makes the chemical a useful research tool for the Handelsman lab. When seedling emergence is significantly higher with metalaxyl treatment than in the control, researchers can conclude with some confidence that the pathogen targeted by metalaxyl is affecting the control.

It is by establishing a standard of comparison that the agrichemical industry influences research on UW85. To point to the pervasive or systemic character of this influence—its effect on experimental design, presentation of research results, the commercial potential of UW85, and available research tools—is not to suggest that the impact of industry in this instance has been unambiguously negative. Indeed, existing research on metalaxyl provides lab workers with purchase on what diseases UW85 controls.

At the same time, the discovery of how UW85 works is less impressive, at some level, if its performance is not comparable or superior to metalaxyl's. Even without direct industry involvement, the commercial potential of UW85 will be shaped by how the cost and field performance of this biocontrol agent compare to traditional chemical pesticides. Had research and development on chemical pesticides never been undertaken, such a contrast could not be made. Instead, the disease control potential of UW85 might be measured against farmers' "cultural practices" (e.g., crop rotation) or a no–treatment option.

There is another way to understand the indirect constraints on research practice that result from industry support for a particular research trajectory. Industry can shape research by not supporting it as well as by supporting it. According to one source, "biological control has never achieved anything close to its potential. A decade ago, proponents forecast that natural products would capture half of all pesticide sales. Yet companies found it expensive and time-consuming to develop biopesticides that worked as fast and as cost-effectively as synthetics" (Groves 1998, D7). Under these circumstances, industry did not support a great deal of biocontrol research, and lack of public funding has been equally responsible for the failure of biocontrol research to live up to its so-called potential. According to a recent report, "[f]ewer than 10 percent of the 1,800 research scientists at the USDA's Agricultural Research Service—and less than 10 percent of the service's $745 million research budget—are focused on developing natural crop-protection methods" (Groves 1998, D7). These effects are indirect but inescapable.

The indirect effects of resource allocation can be seen in many other cases of agricultural research. The research on hybrid seed development directed attention away from the study of improved open-pollinated varieties (Kloppenburg 1988). Focus on research on engineering plants with genes that code for herbicide or disease resistance or with the so-called "terminator" gene may be diverting atten-tion from research on more sustainably oriented and organic agriculture.

It is difficult to say how common the effects of the historical dominance of a research field by industry are in shaping university science. Work by histo-rians of science suggests that this case is not unique. Sawyer (1996), for exam-ple, shows how the measures of success of biocontrol agents studied by citrus fruit researchers in California were affected by cosmetic standards made possi-ble by industry development and widespread adoption of chemical pesticides in citrus agriculture. Similarly, recent interest in intensive use of *Bacillus thuringien-sis* (Bt) in agriculture comes from food industry demand for "unblemished pro-duce at any cost" (Jenkins 1998, 16). With an unblemished produce standard, researchers' efforts are likely to be directed to the insertion of relevant Bt genes into plants, instead of the judicious use of this microorganism in the interest of extending its agriculturally useful life. Whatever the effect on university research of industry dominance of particular fields of investigation, understanding and examining these effects ought not to be left to historians alone. They are surely important matters for public policy consideration.

## CONCLUSION

Insofar as the characteristics of the laboratory I studied are common in academic research settings, my experience as a participant observer in Professor Handelsman's lab suggests that to date the debate on UIRs has overlooked areas that merit broad public discussion. We have focused on the direct effects of industry involvement in academic research. Researchers and policy analysts have looked at contractual relations between science-based companies and university researchers. We have examined what these relations mean for the free flow of information. Investigators have assessed the impact of industry funding of academic research on university research agendas.

My experience in the lab introduced me to a highly reflective group of scientists who wish to work in the public interest. Yet even this group cannot escape the systemic and indirect influence of a world dominated by a particular set of values concerning intellectual property protection and the historical role of a specific industry in a particular area of scientific research. The laboratory cannot escape broad social assumptions concerning patenting. These are deeply institutionalized myths. Indeed, as things currently stand, if they wish to commercialize their inventions, lab workers must patent them. In addition, the Handelsman laboratory cannot escape the influence of chemical industry dominance of fungicide development. The industry standard is an "obligatory point of passage" for their work.

There are a variety of administrative and policy responses that might flow from this kind of analysis. First, consideration should be given to institutionalizing a kind of social reflexivity. That is, universities and governments should consider establishing mechanisms that will allow regular review of university-industry relationships to determine not only whether secrecy is increasing and scholarly research agendas are being skewed by industry sponsorship, but also to determine whether intellectual property protection enhances university scientists' control of their research and whether the time spent seeking this protection is offset by the financial remuneration which results from commercial licensing.

Second, the uncertain social benefit of unexamined and sometimes indiscriminate business efforts to create intellectual property should lead state and federal governments to support research on the economic benefits and costs of intellectual property protection and to experiment with economic incentives for companies and individual researchers who keep their inventions in the public domain.

In addition, resources and energy ought to be allocated to experiments with serious alternatives to traditional intellectual property protection arrangements. For example, Schulman (1999, 188, 189) proposes a practice worth careful consideration: "conceptual zoning." As Schulman notes, traditional land use zoning is used to protect communities from undesirable acts by individual owners. In the case of "conceptual zoning," compulsory licensing of certain kinds of patents could be required and exemptions from exclusive ownership rights could be mandated "in cases where the desired knowledge is sought for noncommercial purposes" (189).

Finally, national and state fiscal difficulties notwithstanding, policymakers should consider establishing funds for university research that promises socially beneficial applications, but where, in the short term, these applications will not be able to compete with commercially developed products.[6]

The indirect effects of the commercial environment on university science are harder to uncover and more difficult to compensate for at a policy level than the direct impacts which have been the focus of attention in discussions of UIRs, and they are easier to ignore. However, although such indirect and pervasive effects may be part and parcel of a system predicated on private property and profit—a system which will be with us for the foreseeable future—it is only by attending to such effects that we can craft policies that intelligently balance equity and efficiency while preserving core democratic values.

## NOTES

The author would like to thank Professor Jo Handelsman and her lab members Elizabeth Blackson, Mark Bittinger, Lynn Jacobson, Laurie Luther, Jocelyn Milner, Eduardo Robleto, Eric Stabb, Sandy Stewart, and Liz Stohl for making this study possible. Other scientists who were helpful include: Professor Robert Goodman, Scott Bintrim, and Kevin Smith. For their useful comments on earlier versions of this paper, the author would like to thank Elizabeth Blackson, Jennifer Croissant, Scott Frickle, Jo Handelsman, Jack Kloppenburg, and Steve Vallas. Research support for the study of which this paper forms a part was provided by the School of History, Technology, and Society at the Georgia Institute of Technology, the Georgia Tech Foundation, and the United States National Endowment for the Humanities. This paper is a revised version of Daniel Lee Kleinman (1998), "Pervasive Influence: Intellectual Property, Industrial History, and University Science," *Science and Public Policy* 25(2):95–102. I thank Bill Page, publisher of *Science and Public Policy*, for permission to publish a revised version of that paper in this volume.

1. In addition to the research I discuss in this section, see Louis and Anderson (1998) and Shenk (1999) for very recent examples of work attentive to immediate and direct effects.

2. I am, of course, not alone in drawing attention to the way the larger environment in which university science is undertaken affects the practices of scholars. See

especially Hackett's chapter in this volume and the work by Slaughter and Rhoades (1990, 1993, 1996; Rhoades and Slaughter 1997; Slaughter and Leslie 1997).

3. Handelsman stresses that although WARF's "success rate" is low, the organization's patenting and related investment efforts provide a substantial sum to the University of Wisconsin, between about $16 and $18 million annually.

4. I use Latour's term *obligatory point of passage* with some reservation, as I believe his analysis is vastly too agency-centered; he focuses too much attention on the ways "actor networks" and "obligatory points of passage" can be enabling and not enough attention on the ways scientists' practices are constrained. Still the metaphor conjures a useful image.

For an example of Latour's enthusiastic use of the concept 'obligatory point of passage,' see Latour (1988, 43–49). For my critique of agency-centered analysis in science studies, see Kleinman (1998).

5. There are plant diseases for which biocontrol agents constitute the only treatment. In these cases, obviously, effectiveness of the biocontrol agent would not be measured against a chemical treatment.

6. Research on biocontrol agents might conceivably fit this characterization.

# 11

# The Gloves Come Off: Shattered Alliances
# in Science and Technology Studies

## LANGDON WINNER

The acrimonious disputes surrounding social studies of science today reflect long-standing disagreements about the character and purpose of inquiry in this field. The publication of Higher Superstition underscores how nasty these quarrels can be, perhaps foreshadowing explosive clashes between the two cultures in years to come (Gross and Levitt 1994). One might have hoped spirits less malicious than Gross and Levitt's would have been the ones to bring these conflicts to light. But for those who have followed the development of science and technology studies (STS) over the years, it has been obvious that eventually the other shoe would drop, that someday it would occur to scientists and technologists to ask: why do the descriptions of our enterprise offered by social scientists and humanists differ so greatly from ones we ourselves prefer? How much longer should we put up with this?

In my experience, four basic projects have inspired research and thinking in STS during the past several decades. While these projects are by no means mutually exclusive, they do serve as food for fairly distinct groups of interests. Within the sprawling interdisciplinary community of STS in America and Europe, one finds widely different intellectual approaches, different expectations about the value of results, and different understandings about the audiences that will ultimately judge and sponsor such work. Tensions between these projects and shifting alliances between them account for much of the vitality of STS, as well as for the flare-ups we see at present.[1]

In the first project, the great challenge is simply to understand how modern science and technology work, how their various practices, institutions, and tangible products have developed. Success in this project comes in providing satisfactory explanations. Thus, one might ask: exactly how did molecular biology arise? What contexts and influences contributed to its rise? How did laboratories take shape? How did career patterns in biology change? What were the roles of government and of business in the rise of this research? Inquiries of this kind have flourished largely because it has never been common for working scientists and technical professionals to write their own histories or to explore the social dynamics of their work. Filling this gap, scholarship in the mode of historical and social explanation has expanded steadily in the United States and Europe during the past thirty years. What was once a dearth of writing in this area has been replaced by a flood of books and journal articles.

Many of those who have embraced STS in its purely explanatory mode have come to the subject from backgrounds in the natural sciences and engineering. Having once made professional commitments to the fields they now study as research objects, it is not surprising that STS scholars of this stripe often have extremely sanguine views of what science and technology are all about. Contrary to what Gross and Levitt assert, it is fairly common among historians in STS to believe that the results of their research should pass muster with colleagues in their former professions and that findings should cast a favorable light on the professions under study. By the same token, it is by no means uncommon for scholars of this stripe to view criticism of science and technology with disdain. The conclusion that STS is a hotbed of critical views simply disregards the extremely conventional but solid scholarship that comprises much historical and sociological research in this area.

For thinkers in the second project, science and technology studies presents a convenient domain in which to develop conceptual models and approaches taken from the home disciplines in the humanities and social sciences. For a number of reasons, none of the fields of human studies organized during the academic reforms of the late nineteenth century chose to focus upon science and technology as such. Intellectual boundaries that seemed sensible at the time meant that the tools of history, sociology, anthropology, political science, literary criticism, and others would not be used to study the inner workings of the natural sciences or engineering. It was widely assumed that these pursuits were not amenable to the same methods used to study politics, world history, kinship relations, or modern fiction. In the decades after World War II, however, it became obvious that because many of the transformations in modern life were crucially connected to developments in science and technology, the social sciences would need to prove their origins and effects or risk being completely out of touch. Hence the rise of a host of new disciplinary and interdisciplinary efforts that by the late 1960s were asking: why not create a true sociology of science? Why not apply the tools of literary criticism to scientific communication?

Research in this vein has spawned a new cottage industry, applying a wide range of concepts and approaches from the humanities and social sciences to the study of science and technology. Much of this work has drawn upon the rise of poststructuralist, and postmodernist theoretical strategies, insisting that the discursive practices of scientific and technical fields were ripe for deconstruction. Unlike scholars coming from backgrounds in science and engineering, however, those who have arrived from the humanities and social sciences are little worried about how their findings will be received by the people and institutions they study. Characteristic of such work has been the spirit of imaginative interpretation and unfettered speculation that has enlivened other areas in which the new cultural studies have taken root. If the results appear critical of contemporary science and technology, so much the better, advocates of the new scholarship are inclined to think; it is simply an indication that the probes have begun to touch raw nerves.

For those engaged in yet a third project, the attraction of STS involves the need to respond to a host of practical problems that have arisen in the world as a consequence of changing scientific knowledge and various technological applications. Here the focus is not so much that of explaining the development of science and technology or showing how new intellectual tools can be applied, but rather analyzing specific troublesome issues, for example, the safety of nuclear power plants, environmental devastation caused by overdevelopment, hazards to public health and safety caused by the dumping of toxic wastes, and so on. While research of this kind aspires to both intellectual richness and policy relevance, what matters in the end are proposals to remedy the problems described. For that reason, advocates of this project are closely oriented toward decision makers in Washington, D.C., and social movements that might effect positive change.

Concerns of this sort may have little connection to the two projects mentioned above. Those at work on problem-centered research may find little value in writing about closely related topics in the history, sociology, or cultural studies of science. In fact, it is fairly common for them to lament that STS is dissolving in a swamp of arcana—studies of laboratory life and elaborate interpretations of material artifacts, for example—while the real troubles of the world go begging. After all, what good is all this novel theorizing if it does not generate practical remedies?

A fourth collection of concerns in STS attracts philosophers and social theorists. Here the focus turns to what many thinkers have argued is a profound crisis in the underlying conditions of modern life and thought. The development of modernity has gone badly wrong, not only at the level of specific, vexing social problems but in its fundamental core of ideas and institutions, especially those that involve science and technology. While attempts to fathom the nature of the crisis vary from writer to writer—from Marx to Mumford, from Heidegger to Ellul, from Habermas to Foucault—the point of inquiry is to locate philosophical, historical, and cultural origins of phenomena closer to hand.

In its very nature, research of this kind is both radical and critical; it seeks to look deeper, to probe what may be highly general sources of contemporary discrimination and to suggest change of the most fundamental kind. This does not mean, as opponents sometimes claim, that thinking in this key is essentially "antiscience" or "antitechnology." But it certainly does mean that prevailing ideas and practices in science and technology must be subjected to close scrutiny with no a priori declarations of endorsement. The fact that STS in a philosophical vein resists collegial appeals to join the chorus affirming science as a grand and glorious enterprise has often attracted the wrath of science boosters. No less a man than Lewis Mumford became the focus of such ire when Gerald Holton, his erstwhile friend and one of today's most Quixote-like crusaders against "antiscience," concluded that Mumford's Pentagon of Power was simply an attack on scientific rationality, the thesis of Holton's bitter review in the New York Times.[2]

Roughly speaking then, one can locate the purposes of much STS research and teaching within the four projects I have summarized. My own work, for example, flows primarily from two of these: expressing a desire to confront what I perceive to be a systematic disorder in modern life, a disorder manifest in technology-centered ways of living that I regard as unfriendly to any sane aspiration for human being; and applying concepts and approaches of a particular discipline, political theory, to questions about the significance of technology for political life.

Whatever one's personal project, however, all who enter STS must navigate an intellectual terrain characterized by diverse intentions and uneasy alliances. And always just offstage are sources of support and opposition presented by those who have a strong stake in where the discussion moves. As I entered STS in the middle 1970s, the desire to explain the inner workings of science and technology was clearly the one most prominent among my colleagues. It was also the purpose most resonant with university scientists, engineers, and administrators who clearly hoped that better biographies of scientists and inventors, and better histories of the various fields of research and development, would shine a favorable light on their professions. They looked to STS to contribute to the celebration of science and technology as crucial to the progress of civilization, describing in detail how the great wealth of knowledge and social benefit had been achieved.

At the time it appeared that a second important STS project, that of extending the conceptual equipment of the humanities and social sciences into this new domain, would fit the program of hagiography and celebration very nicely. Support for the new research would demonstrate that the long-neglected human dimensions were finally receiving emphasis. When the results came in, they would reveal that what sometimes appears to be hard-nosed, even soulless work by scientists and engineers is, in fact, morally complex, aesthetically deep, and culturally refined. Several of the STS programs founded at colleges and universities during this period upheld science appreciation as their central educational aim, one that many faculty members enthusiastically endorsed. Such initiatives, in my view, could well have been named HSTS: Hooray for Science, Technology, and Society.

Other themes in STS research, however, were far more difficult for scientists, engineers, and university bureaucrats to stomach. A growing list of studies on social and environmental ills accompanied by a hefty stack of philosophical critiques of technological society sometimes made it seem that the scientific and technical professions were being unjustly attacked, indiscriminately blamed for all the ills of modern society. In that mood, those who studied particular signs of social stress and ecological decay were often accused of having negative and unbalanced views of science and technology. Those who studied the philosophical roots of systems analysis, artificial intelligence, and the like were quickly labeled antitechnology malefactors. Thus a senior scientist wellknown for his accomplishments in both computing and social criticism was repeatedly rebuffed in his attempts to join the STS program at MIT, his home institution. Persisting in his lively exposé of the follies of instrumental rationality, the man became persona non grata among the timid souls who ran the program and who sought to please the institute's faculty by singing the praises of science and engineering.

Conflicting views about the proper aims of STS occasionally erupted into nasty struggles. Some notable tenure battles, especially David F. Noble's lawsuit against MIT, gave adequate evidence that the field was not one big happy family. Indeed, when scholars with critical views have joined university programs geared to celebrating the progress of science and technology, they have often been headed for trouble, not the least of which is opposition from colleagues in the "hard" disciplines. As if to suggest a bold departure, Gross and Levitt propose that scientists respond to antiscience bias by taking control of the tenure and promotion of their colleagues in the humanities. Those who have been around this track before might ask: so what else is new?

Until Higher Superstition broke the silence, the opinions of mainstream scientists about the appropriate boundaries for science and technology studies were seldom openly discussed. Humanists and social scientists were supposed to read subtle messages that informed them of their subservience and act prudently. That is why I was astonished by the frank confession offered by one authoritative spokesman in the early 1980s. The occasion was the meeting of a review panel for the program on Ethics and Values in Science and Technology (EVIST) of the National Science Foundation. Our task was to sift through several dozen research proposals and decide which ones deserved funding. As we began the day's work, an upper-level NSF administrator came into the room to explain the real reason we were there. He noted that EVIST had gained support within the foundation because it helped the scientific community respond to political pressures from Congress and the general public. As problems in ethics and values of science arose in matters of, say, nuclear power or molecular biology, scientists could respond that qualified experts in the humanities and social sciences were looking into the matter. The public's misgivings about problems associated with science and technology could be answered by pointing to research programs in STS where all the social and ethical perplexities, all those knotty "values" questions were being rigorously investigated.

The contrast in purposes had never been so vividly apparent to me. What some of us in STS saw as a bold new intellectual enterprise aimed at challenging prevailing notions about the workings of science and technology was understood by science administrators as a way to blunt or co-opt signs of discontent that had recently erupted in American society. The role of STS scholarship was, in effect, to help defuse public anger by transforming disruptive protests into tame academic pursuits. As the administrator droned on about this strategy, I remember thinking to myself, "Well, it's nice to know what business you're in, at least in the eyes of the people paying the bills."

As background for understanding today's worries about the character of STS, the musings of the NSF administrator must be seen in historical context. As David Dickson has noted, the late 1970s and early 1980s (the years of the Carter and Reagan presidencies) were years that witnessed a shift in relationships between science and society (Dickson 1984). During the previous decade, scientists found themselves subject to pressures to orient research toward national priorities in health care, environmental clean-up, and energy research. Many scientists came to believe that the public's influence on R&D had grown too large, that the direction of science by political policymakers had gotten out of hand. Dickson argues that scientists, galled by what they regarded as excessive democratic control of research agendas, were more than willing to form alliances with other sources of social control. Hence, during the Reagan era scientists supported a turn away from research agendas shaped by a sense of social need toward R&D geared to the ongoing military buildup and quest for "national competitiveness" expressed in the priorities of business firms.

In that setting, anything that could be done to diminish public concerns about social and environmental issues that had surfaced in the 1960s and early 1970s was regarded as a positive step. In Dickson's account, the continuing support for programs like EVIST and the Office of Technology Assessment were among the means chosen by leaders in the scientific and technical community to deflect the tendency of a restless public to assert its role in setting research priorities. Many in STS at the time recognized these pressures and fought them by trying to strengthen ties to social movements and finding ways to address political issues in STS through the popular media. On the whole, however, the development of science and technology studies in the 1980s flowed with the depoliticizing tides of that decade. As reflected in the yearly meetings of the Society for Social Studies of Science, many scholars were content to discuss such things as the social construction of the bicycle during the late nineteenth century rather than, say, the rapid dismantling of industrial workplaces shattering the lives of so many of their contemporaries.

During the past decade the second project on my list has matured, engaging new voices from a wide variety of backgrounds. Once counted on to lend an air of polite refinement to literary and cultural discussions about science, the development of methods in the humanities and social sciences had instead produced a direct challenge to understandings about scientific knowledge and technical application that has been dominant for the better part of two centuries. Feminists, postmodernists, and critical theorists of many persuasions have converged on the idea that, yes, the approaches of the revitalized humanities can well be directed to interpreting what science and technology are all about. In this way of seeing, the descriptions and accounts of scientists and engineers provide useful data, but they are by no means the last word on the matter. This is the feature of the new scholarship that drives Gross and Levitt nuts, because it seems to undermine the authority of science. It is also a feature that has now begun to attract the attention of right-wing militants worried about politically correct speech, multiculturalism, and other supposed dangers lurking in the academy. Indeed, it is not unthinkable that an attack on antiscience could be incorporated into later goose steps of the Contract with America.

The surprise is, of course, that the turn in the 1980s toward abstract, politically disengaged research, focused on projects of explanation and interpretation, eventually produced results that now seem threatening. As researchers in the social sciences and humanities devise increasingly sophisticated ways of asking what's really going on in the development of scientific knowledge and what actually occurs in the construction of technology-centered practices, their answers are sometimes unsettling to those with a supposedly infallible insider's view. A consistent finding in recent STS writings has been that the phenomenon of power infuses science and technology through and through. As scientists and engineers read the books and articles that argue this point, some recoil in utter horror. Arguments of this kind do not match their belief that science is a sure-handed method for obtaining beautiful, objective truths. Neither do they match the central ideal of engineering that technological development is a neutral process that simply pursues the best instrumental results. As they puzzle over the new scholarship, people in the "hard" disciplines sometimes conclude that all the power relations said to exist at the heart of science and technology must have been imported from outside, probably from the social scientists' own mistaken models.

It is too soon to tell whether this irate response will be the one that prevails. What is clear is that the gentlemanly social contract that previously linked STS research to allies in the scientific and technical communities is now under fire. Scientists and engineers have been prepared to abide some degree of criticism coming from problem-oriented and philosophical wings of STS, as long as what they assumed to be the central purpose—production of histories and social analyses portraying science and technology as grand and glorious endeavors—would still be the heart of the enterprise. Will they still be prepared to support STS when it appears to them that the entire field of inquiry has taken off in perverse directions? Perhaps not. Rather than come to terms with the new ideas and enter dialogue with those who advance them, leaders in the scientific and technical communities may simply decide it is best to shut the whole thing down. That is, of course, the recommendation of Higher Superstition, one that ironically confirms much of what STS scholars have been saying. You claim that science is mainly about power? We'll show you! We'll smash this heresy by taking away your jobs and funding. Quod erat demonstrandum.

As before, it is important to notice the context in which these tensions arise. In the 1990s the basic relations between the scientific and technical professions and the rest of society are once again being renegotiated. Concerns of the previous decade about productivity, national competitiveness, and their promised link to the well-being of the whole American populace have been supplanted by a more bracing recognition of what the global economy demands. The idea that the advance of science and technology must contribute to the prosperity of any particular nation has given way to a recognition that entrepreneurs will create high-tech enterprises wherever in the world it is most advantageous to do so. Thus scientists and engineers, like everyone else, must be swift enough and flexible enough to satisfy rapidly changing corporate demand. This means that specific needs of national economies and populations must be deemphasized in favor of priorities established by "the global market." Responding to this situation, educators in science, engineering, and business now commonly warn their students that they must regard themselves as transnational actors and shape their careers accordingly. In a similar vein, government R&D programs formerly aimed at fostering new products for American industries and jobs for American workers are dismantled in favor of letting "the market decide." Preferred now is basic research without reference to any well-defined social need, research that will, in the fullness of time, contribute to economic growth—somewhere in the world.

In this new atmosphere, any variety of discourse that helps remind people of the social character of science and technology may seem counterproductive to those who control the flow of resources. Even once-cherished themes of productivity and competitiveness must be de-emphasized because they may suggest to everyday people that improvements of that kind might actually benefit them. In fact, the scientific advances, technological innovations, and productivity gains in the United States of the past twenty years have gone hand-in-hand with rapid erosion of the real wages of working-class and middle-class Americans. To an increasing extent, the science-based, high-tech economy has brought a widening gap between the extremely well-to-do and the rest of society. By almost any conventional measure, progress, seen as the contribution of advancing science and technology to the living conditions of most people, has been stalled for a long while. For those seeking new roles in the corridors of transnational capital, it may seem undesirable to ask the general populace to ponder how science and technology work nowadays, or whether changes taking place in that sphere are truly beneficial to them. Under the circumstances, it is no surprise that we see the resurgence of the myth of pure, objective science along with the myth of unfettered technological progress, the STS equivalent of "family values."

Disturbing evidence on this score comes from recent episodes of ideological cleansing at the Smithsonian Institution. To mark the fiftieth anniversary of the dropping of the atomic bomb on Hiroshima, the Air and Space Museum planned to show the Enola Gay along with photos and historical descriptions of related events before and after the blast. But the exhibit was canceled with the American Legion and Air Force Association complained that it would show, of all things, the effects of the explosion on its human victims. According to the veterans, such evidence would soil the patriotic message the commemoration ought to convey, a message of strength, victory, and American lives saved by the bomb. Responding to these pressures, museum officials canceled the original exhibit and replaced it with a vastly scaled-down, inoffensive version.[3]

In a similar episode, the broad-ranging $5.5 million exhibition Science in American Life, which opened in April 1994, drew fire from physicists who alleged that it showed too many negative aspects of science and too few positive ones. A letter of complaint sent to the Smithsonian by Burton Richter, president of the American Physical Society, charged that the exhibit gave "a portrayal of science that trivializes its accomplishments and exaggerates any negative consequences. We are concerned that the presentation is seriously misleading, and will inhibit the American public's ability to make informed decisions on the future uses of science and technology" (Kleiner 1995, 42). Among the features of the exhibit that the physicists found offensive were those that emphasized links between scientific research and commercial interests (how unseemly!) and those pointing to the underrepresentation of women and minorities in science (telltale signs of political correctness). The physicists also took umbrage at the barbed-wire fences included in depictions of the Manhattan Project, which built the first atomic bomb.

Eventually, the American Chemical Society (ACS), financial backer of the exhibit, joined the physicists in demanding that the exhibit be cleansed of its distressing messages. Paul Walter, chairman of the ACS, wrote to Smithsonian officials demanding that the exhibit be changed because parts of it "demonstrate a strong built-in tendency to revise and rewrite history in a 'politically correct' fashion" (Macilwain, 1995a, Now Chemists Hit . . . 752). Marvin Lang, chemistry professor at the University of Wisconsin and member of the original advisory board for *Science in American Life*, charged that planning for the show was too greatly influenced by "social scientists and pseudo-scientists who had no idea of how science worked" (Macilwain, 1995b, Smithsonian Heeds . . . 307). Yielding to these pressures, the Smithsonian secretary Michael Heyman agreed to set up procedures to rid the exhibit of its supposedly antiscience bias.

The beleaguered Smithsonian curators who have become a cropper of angry patrons are, of course, members of the STS community. As represented in the halls of the Smithsonian, their work is by no means on the outer edge of science theory but instead expresses some straightforward lessons about the social nature of science and technology, lessons that STS scholars have documented so thoroughly they now seem common sense. That even these mundane themes now seem reprehensible to authority figures who oversee the reputation of science and technology bodes ill for the future relationship between STS and its previous sources of institutional support.

The *modus vivendi* between humanists and social scientists and scientific and technical professionals has fostered the development of challenging new perspectives on both the inner workings and context in which science and technology operate. Apart from occasional stresses, strains, and squabbles, it long seemed that a community of interests would cohere. Now, however, it seems possible that frictions within this coalition, intensified by political upheavals in society at large, may tear long-standing alliances to shreds. Rather than abide (much less sponsor) challenging, advanced work in STS, some in the scientific community yearn to return to the good old days, the days of Carnap, Bronowski, and "our friend the atom."

Members of the broad, diverse networks of STS, including a good many sympathetic scientists and engineers, have yet to respond effectively to this disturbing turn of events. The time for organizing that response may be short.

Self-anointed "pro-science" forces are now forging powerful alliances with radical reactionaries, ones marching in the forefront of every backward step.

## NOTES

1. For portraits of topics and approaches in STS, see Sheila Jasanoff et al., eds., Handbook of Science and Technology Studies (Thousand Oaks, Calif.: Sage 1995). For an overview from fifteen years earlier see Paul T. Durbin, ed., A Guide to the Culture of Science, Technology, and Medicine (New York: Free Press 1980).

2. For details of this sad story see Donald L. Miller, Lewis Mumford: A Life (New York: Weidenfeld and Nicolson, 1989):534–37.

3. For a report on the Enola Gay controversy, see Helen Gavaghan, "Smithsonian to Study Museum's Role after Dropping A-Bomb Exhibition," Nature, 2 February 1995, 371. Historian of science Stanley Goldberg, a member of the Enola Gay Exhibit Advisory Board who resigned in protect, reflects on the significance of the debacle in "Smithsonian Suffers Legionnaires' Disease," Bulletin of the Atomic Scientists 5 (May/June 1995):28–33. Looking back upon successful efforts by veterans' groups and Congress "to censor the conclusions of sound historical scholarship," Goldberg warns, "That kind of thought control should have no place in a government committed to democracy. I believed that issue had been settled in the 1950s, when McCarthyism was laid to rest. Apparently I was wrong" (33).

# Literature Cited

Abbott, A. 1988. *The system of professions: An essay on the division of expert labour.* Chicago: University of Chicago Press.

Albert, M. B., D. Avery, F. Narin, and P. McAllister. 1991. Direct validation of citation counts as indicators of industrially important patents. *Research Policy* 20:251–259.

Aldrich, H., and J. Pfeffer. 1976. Environments of organizations. *Annual Review of Sociology* Vol. 2:79–105.

Allison, P. D., and J. A. Stewart. 1974. Productivity differences among scientists: Evidence for accumulative advantage. *American Sociological Review* 39:596–606.

American Association of University Professors (AAUP). 1983. Academic freedom and tenure: Corporate funding of academic research. *Academe* (November–December):18a–23a.

Anderson, K. L. 1985. College characteristics and change in students' occupational values. *Work and Occupations* (August) 12:307–328.

Argyres, N. S., and J. P. Liebeskind. 1998. Privatizing the intellectual commons: Universities and the commercialization of biotechnology. *Journal of Economic Behavior and Organization.* 35(4):427–454.

Arrow, K. J. 1987. Reflections on the essays. In *Arrow and the Ascent of Modern Economic Theory*, ed. G. R. Feiwel, 659–689. New York: New York University Press.

Ashford, N. A. 1983. A framework for examining the effects of industrial funding on academic freedom and the integrity of the university. *Science, Technology, & Human Values* 8:16–23.

Association of University Technology Managers. 1997. *Licensing survey, FY 1997: Executive summary.* The survey is only available by purchase through The Association of University Technology Managers, Inc.

Astin, A. W. 1977. *Four critical years.* San Francisco: Jossey-Bass.

Bailyn, L. 1980. *Living with technology: Issues at mid-career.* Cambridge, MA: MIT Press.

Baldwin, D. R. 1988. Academia's new role in technology transfer and economic development. *Research Management Review* 2(2):1–16.

Barney, J. B., and E. J. Zajac. 1994. Competitive organizational behavior: Toward an organizationally-based theory of competitive advantage. *Strategic Management Journal* 15:5–9.

Bayles, M. D. 1981. *Professional ethics*. Belmont, CA: Wadsworth Publishing Co.

Ben-David, J. 1977. *Centers of learning, Britain, France, Germany, and the United States*. New York: McGraw-Hill.

Bentley, R., and R. Blackburn. 1990. Changes in academic research over time: A study of institutional accumulative advantage. *Research in Higher Education*. 31(4):327–353.

Bereano, P. L. 1986. Making knowledge a commodity: Increased corporate influence on universities. *IEEE Technology and Society Magazine* (December):8–17.

Bijker, W. 1995. *Of bicycles, bakelites, and bulbs: Toward a theory of sociotechnical Change*. Cambridge: MIT Press.

Bijker, W., T. P. Hughes, and T. J. Pinch, eds. 1987. *The social construction of technological systems*. Cambridge: MIT Press.

Blumenthal, D. 1992. Academic-industry relations in the life sciences: Extent, consequences, and management. *Journal of the American Medical Association*. 268(23):3344–349.

Blumenthal, D. 1996. Ethical issues in academic-industry relationships in the life sciences: The continuing debate. *Academic Medicine* 71:1291.

Blumenthal, D., N. Causino, E. Campbell, and K. S. Louis. 1996. Relationships between academic institutions and industry in the Life Sciences—An industry survey. *New England Journal of Medicine* 334(6):368–373.

Blumenthal, D., S. Epstein, and J. Maxwell. 1986. Commercializing university research: Lessons from the experience of the Wisconsin Alumni Research Foundation. *The New England Journal of Medicine* 314:1621–1626.

Blumenthal, D., M. Gluck, K. Louis, and D. Wise. 1986. Industrial support of university research in biot echnology. *Science* 231:242–246.

Blumenthal, D., M. Gluck, K. Louis, M. Sonte, and D. Wise. 1986. University-industry research relationships in biotechnology: Implications for the university. *Science* 232:1361–1366.

Bonitz, M., E. Bruckner, and A. Scharnhorst. 1997. Characteristics and impact of the Matthew Effect for countries' scientometrics. *Scientometrics* 40(3):407–422.

Bourdieu, P., and J.-C. Passeron. 1977. *Reproduction in education, society, and culture*. London: Sage.

Bowie, N. E. 1994. *University-business partnerships: An assessment*. Laynham, MD: Rowman & Littlefield Publishers, Inc.

Boyle, J. 1996. *Shamans, software, and spleens: Law and the construction of the information society*. Cambridge: Harvard University Press.

Braun, H.-J. 1992. Introduction. Symposium on 'Failed innovations'. *Social Studies of Science* 22:213–230.

Braverman, H. 1974. *Labor and monopoly capital: The degradation of work in the twentieth century.* New York: Monthly Review Press.

Brett, A. M., D. V. Gibson, and R. W. Smilor, eds. 1991. *University spin-off companies: Economic development, faculty entrepreneurs, and technology transfer.* Savage, MD: Rowman and Littlefield Publishers.

Brown, J. S., and P. Duguid. 1991. Organizational learning and communities of practice: Toward a unified view of working, learning, and innovation. *Organization Science* 2:40–57.

Bucciarelli, L. L. 1994. *Designing engineers.* Cambridge: MIT Press.

Burns, T., and G. M. Stalker. 1961. *The management of innovation.* London: Tavistock Publications.

Busch, L., and W. B. Lacy. 1993. *Science, agriculture, and the politics of research.* Boulder, CO: Westview Press.

Bush, V. 1945. *Science, the endless frontier: A report to the President by Vannevar Bush, director of the Office of Scientific Research and Development.* Washington, DC: Government Printing Office.

———. 1946. *Endless horizons.* Washington, DC: Public Affairs Press.

Buttel, F., and J. Belsky. 1987. Biotechnology, plant breeding, and intellectual property: Social and ethical dimensions. *Science, Technology, and Human Values* 12:31–49.

Byerly, R., and R. A. Pielke. 1995. The changing ecology of United States science. *Science* 269(5230):1531–1532.

———. 1998. Beyond basic and applied. *Physics Today* 51(2):42–46.

Callon, M. 1986a. Some elements of a sociology of translation: Domestication of the scallops and the fisherman of Saint Brieuc Bay. In *Power, action, belief: A new sociology of knowledge?* ed. J. Law, 196–233. London: Routledge and Kegan Paul.

———. 1986b. The sociology of an actor-network: The case of the electric vehicle. In *Mapping the dynamics of science and technology,* ed. M. Callon, J. Law, and A. Rip, 19–34. Basingstoke, UK: Macmillan.

———. 1994. Is science a public good? *Science, Technology, and Human Values* 19(3):395–424.

Campbell, T. 1997. Public policy for the 21st century: Addressing potential conflicts in university-industry collaboration. *The Review of Higher Education* 20(4):357–379.

Carroll, G. R., and M. T. Hannan. 1989a. Density dependence in the evolution of populations of newspaper organizations. *American Sociological Review* 54:524–541.

———. 1989b. Reply to Zucker. *American Sociological Review* 54:545–548.

Chubin, D. How large an R&D enterprise? In *The fragile contract: University science and the federal government,* ed. D. H. Guston and K. Keniston, 118–144. Cambridge: MIT Press.

Chubin, D. E., and E. J. Hackett. 1990. *Peerless science: Peer review in U.S. science policy.* Albany, NY: State University of New York Press.

Clark, B. R. 1983a. *Values in higher education: Conflict and accommodation.* Tucson, AZ: Center for the Study of Higher Education.

———. 1983b. *The higher education system: Academic organization in cross-national perspective.* Berkeley: University of California Press.

———. ed. 1993. *The research foundations of graduate education.* Berkeley, CA: University of California Press.

———. 1995. *Places of inquiry: Research and advanced education in modern universities.* Berkeley: University of California Press.

Clark, C. 1997. *Radium girls: Women and industrial health reform 1910–1935.* Chapel Hill: University of North Carolina Press.

Clegg, S. 1975. *Power, rule, and domination: A critical and empirical understanding of power in sociological theory and organizational life.* London: Routledge & Kegan Paul.

Coggeshall, P. E., and J. C. Norvell. 1978. Changing postdoctoral career patterns for biomedical scientists. *Science,* 202:487–493.

Cohen, J. 1997. Please pass the data. *Science* 276 (27 June):1961.

Cohen, L. R., and R. G. Noll. 1994. Privatizing public research. *Scientific American* 271(3) (September):72–77.

Cohen, M., and J. G. March. 1974. *Leadership and ambiguity: The American college president.* Boston, MA: Harvard University Press.

Cohen, M., J. March, and J. Olsen. 1972. A garbage can model of organization choice. *Administrative Science Quarterly* 17:1–25.

Cohen, W., and D. Levinthal. 1990. Absorbtive capacity: A new perspective on learning and innovation. *Administrative Science Quarterly* 35:128-152.

Cole, J. R., and S. R. Cole. 1973. *Social stratification in science.* Chicago: University of Chicago Press.

Collins, R. 1979. *The credential society: An historical sociology of education and stratification.* New York: Academic Press.

Collins, H. M. 1990. *Artificial experts: Social knowledge and intelligent machines.* MIT Press, Cambridge.

Committee on Criteria for Federal Support of Research and Development. 1995. *Allocating funds for science and technology.* Washington, D.C.: National Academy Press.

Cook, T., and D. Campbell. 1979. *Quasi-experimentation: Design and analysis for field settings.* Chicago: Rand-McNally.

Cooley, M. 1980. *Architect or bee? The human/technology relationship.* Boston: South End Press.

Cordes, C. 1991. Congress earmarks $493 million for specific universities; Critics deride much of the total as 'pork barrel' spending. *The Chronicle of Higher Education* XXXVII (27 February):A1, A21.

————. 1998. The academic pork barrel begins to fill up again. *The Chronicle of Higher Education* XLIV (19 June):A30–31.

Cowan, R. S. 1985. How the refrigerator got its hum. In *The social shaping of technology*, ed. D. Mackenzie and J. Wajcman, 202–218. Open University Press, Milton Keynes.

Crane, D. 1972. *Invisible colleges: Diffusion of knowledge in scientific communities.* Chicago: University of Chicago Press.

Czujko, R., D. Kleppner, and S. Rice. 1991. Their most productive years: Young physics faculty in 1990. *Physics Today* (February):37–42.

Dasgupta, P., and P. David. 1987. Information disclosure and the economics of science and technology. In *Arrow and the ascent of modern economic theory*, ed. G. R. Feiwel, 519–542. New York: New York University Press.

————. 1994. Toward a new economics of science. *Research Policy* 23(5):487–521.

Daston, L. 1992. Objectivity and the escape from perspective. *Social Studies of Science* 22:597–618.

Davis, M. 1982. Conflict of interest. *Business & Professional Ethics Journal* (Summer):17–27.

de Solla Price, D. 1963. *Little science, big science.* New Haven: Yale University Press.

De George, R. T. 1997. *Academic freedom and tenure, Ethical issues.* Lanham, MD: Rowman & Littlefield.

Derber, C., ed. 1982. *Professionals as workers: Mental labor in advanced capitalism.* Boston: G. K. Hall.

Dey, E. L., J. F. Milem, and J. B. Berger. 1997. Changing patterns of publication productivity: Accumulative advantage or institutional isomorphism? *Sociology of Education* 70:308–23.

Dickson, D. 1984. *The new politics of science.* New York: Pantheon.

————. 1988. *The new politics of science.* Chicago: University of Chicago Press.

Dimaggio, P., and W. W. Powell. 1983. The iron cage revisited: Institutional isomorphism and collective rationality in organizational fields. *American Sociological Review* 48:147–160.

DiMaggio, P. J. 1994. Culture and economy. In *The handbook of economic sociology*, ed. N. J. Smelser and R. Swedberg, 27–57. Princeton NJ: Princeton University Press.

Dooris, M. J. and J. S. Fairweather. 1994. Structure and culture in faculty work: Implications for technology transfer. *The Review of Higher Education* 17(2): 161–177.

Dougherty, D. 1996. Organizing for innovation. In *Handbook of organization studies*, eds. S. R. Clegg, C. Hardy, and W. R. Nord, 424–439. London: Sage Publications.

Downey, G. 1998. *The machine in me: An anthropologist sits among computer engineers.* New York: Routledge.

Dubinskas, F. A. 1988. *Making time: Ethnographies of high-tech organizations*. Philadelphia: Temple University Press.

Durbin, P. T., ed. 1980. *A guide to the culture of science, technology, and medicine*. New York: Free Press.

Dworkin, G. 1987. Commentary: Legal and ethical issues. *Science Technology, and Human Values* 12:63–64.

Eichhorn, R. L. 1969. The student engineer. In *The engineers and the social system*, ed. Robert Perrucci and Joel E. Gerstl, 141–160. New York: Wiley.

Ercolano, V. 1994. Ethical dilemmas: When faculty responsibilities conflict. *ASEE Prism*. 4(2):20–24.

Erikson, K. T. 1976. *Everything in its path*. New York: Simon and Schuster.

Etzkowitz, H. 1983. Entrepreneurial scientists and entrepreneurial universities in American academic science." *Minerva* XXI:198–233.

———. 1989. Entreprenurial science in the academy: A case of transformation of norms. *Social Problems* 36(1):14–29.

———. 1990. The second academic revolution: The role of the research in economic development. In *The R&D system in transition*, ed. Susan E. Cozzens, Peter Healy, Arie Rip, and John Ziman, 109–124. Dordrecht, the Netherlands: Kluwer.

———. 1997. The entrepreneurial university and the emergence of democratic corporatism. In *Universities and the global knowledge economy: A triple helix of university-industry-government relations*, ed. H. Etzkowitz and L. Leydesdorff, 141–152. London: Pinter Press.

Etzkowitz, H., and L. Leydesdorff, eds. 1997. *Universities and the global knowledge economy*. Washington, D.C.: Pinter.

———, and A. Webster. 1995. Science as intellectual property. In *Handbook of science and technology studies*, ed. S. Jasanoff, G. E. Markle, J. C. Petersen and T. Pinch. Thousand Oaks, CA: Sage Publications.

Fairweather, J. 1989. Academic research and instruction: The industrial connection. *Journal of Higher Education* 60(4):82.

Fassin, Y. 1991. Academic ethos vs. business ethics. *International Journal of Technology Management* 6(5/6):533–546.

Feldman, M. P. 1994. *The geography of innovation*. Boston: Kluwer.

Feldman, M. P., and R. Florida. 1994. The geographic sources of innovation. *Annals of the Association of American Geographers*. 84(2):210–229.

Feller, I. 1990. Universities as engines of R&D based economic growth: They think they can. *Research Policy* 19:335–348.

———. 1991. Issues for the HE sector: Lessons from U.S. experiences with collaboration. *Symposium: The true price of collaborative research*. London: Royal Society.

Fligstein, N. 1987. The intraorganizational power struggle: Rise of finance personnel to top leadership in large corporations, 1919–1979. *American Sociological Review* 52:44–58.

Forsythe, D. E. 1996. New bottles, old wine: Hidden cultural assumptions in a computerized explanation system for migraine sufferers. *Medical Anthropology Quarterly* 10(4):551–574.

Friedly, J. 1996. New anticoagulant prompts bad blood between partners. *Science* 271(29):1800–1801.

Friedson, E. 1971. Professions and the occupational principle. In *The professions and their prospects*, ed. E. Friedson. Thousand Oaks, CA: Sage.

Fuchsberg, G. 1989. Prospect of commercial gain from unconfirmed discovery prompted Utah U. officials to skirt usual scientific protocol. *Chronicle of Higher Education* (17 May):A5.

Fujimura, J. H. 1987. Constructing 'do-able' problems in cancer research: Articulating alignment. *Social Studies of Science* 17:257–293

Galison, P., and B. Hevly, eds. 1992. *Big science: The growth of large-scale research.* Stanford, CA: Stanford University Press.

Gardner, P., and A. Broadus. 1990. *Pursuing an engineering degree: An examination of issues pertaining to persistence in engineering.* East Lansing, MI: Collegiate Employment Research Institute, Michigan State University.

Gavaghan, H. 1995. Smithsonian to study museum's role after dropping A-bomb exhibition. *Nature* (February):371.

Geertz, C. 1973. *The Interpretation of Cultures.* New York: Basic Books.

Geiger, R. 1986. *To advance knowledge: The growth of the American research university, 1900–1940.* New York: Oxford University Press.

———. 1988. Milking the sacred cow: Research and the quest for useful knowledge in the American university since 1920. *Science, Technology, and Human Values* 13(3&4):332–348.

Geiger, R. L. 1993. *Research and relevant knowledge: American research universities since World War II.* Oxford: Oxford University Press.

General Accounting Office. *University finances: Research revenues and expenditures* (GAO/RCED-86-162BR). Washington, D.C.: USGAO (July).

Giamatti, A. 1982. The university, industry, and cooperative research. *Science* 218:1278–1280.

Gibbons, M., C. Limoges, H. Nowotny, S. Schwartzman, P. Scott, and M. Trow. 1994. *The new production of knowledge: The dynamics of science and research in contemporary societies.* London: Sage Publications.

Giddens, A. 1971. *Capitalism and modern social theory.* Cambridge, UK, Cambridge University Press.

———. 1984. *The constitution of society: Outline of the theory of structuration.* Berkeley: University of California Press.

Gieryn, T. F. 1999. *Cultural boundaries of science: Credibility on the line*. Chicago: University of Chicago Press.

Goldberg, S. 1995. Smithsonian suffers Legionnaires' Disease. *Bulletin of the Atomic Scientists* 5:28–33.

Goodstein, David L. 1995. After the big crunch. *Wilson Quarterly* (Summer):53–60.

Gordon, G., and S. Marquis. 1966. Freedom, visibility of consequences, and scientific innovation. *American Journal of Sociology* 72:195–202.

Gouldner, A. W. 1958a. Cosmopolitans and locals: Toward an analysis of latent social roles—I. *Administrative Science Quarterly* 2:281–306.

———. 1958b. Cosmopolitans and locals: Toward an analysis of latent social roles—II. *Administrative Science Quarterly* 2:444–480.

Greenwood, E. 1957. Attributes of a profession. *Social Work* (July):45–55.

Griliches, Z. 1990. Patent statistics as economic indicators: A survey. *Journal of Economic Literature.* 28:1661–1707.

Gross, P. R., and N. Levitt. 1994. *Higher superstition: The academic left and its quarrels with science*. Baltimore: Johns Hopkins University Press.

Groves, M. 1998. Working out the bugs. *The Los Angeles Times* (May 4):D1, D7.

Guston, D. 1999. Stabilizing the boundary between US politics and science. *Social Studies of Science* 29(1):87–112.

Guston, D., and K. Keniston, eds. 1994. *The fragile contract: University science and the federal government*. Cambridge, MA: MIT Press.

Hackett, E. J. 1987. Funding and academic research in the life sciences: Results of an exploratory study. *Science and Technology Studies* 5:134–147.

———. Organizational perspectives on university-industry research relations [this volume].

———. Science as a vocation in the 1990s: The changing organization culture of academic science [this volume].

Hagstrom, W. O. 1964. Anomy in scientific communities. *Social Problems* 12:186–95.

———. 1965. *The scientific community*. New York: Basic Books.

———. 1974. Competition in science. *American Sociological Review* 39:1–18.

Hamilton, G. G. 1994. Civilizations and the organization of economies. In *The handbook of economic sociology*, ed. N. J. Smelser and R. Swedberg, 183–205. Princeton, NJ: Princeton University Press.

Handelsman, J., S. Raffel, E. Mester, L. Wunderlich, and C. Grau. 1990. Biological control of damping-off of alfalfa seedlings with *Bacillus cereus* UW85. *Applied and Environmental Microbiology*:713–718.

Hannan, M., and J. Freeman. 1997. The population ecology of organizations. *American Journal of Sociology* 92:929–964.

Haskell, T. L. 1984. Professionalism versus capitalism: R. H. Tawney, Emile Durkheim, and C. S. Peirce on the disinterestedness of professional communities. In *The authority of experts: Studies in history and theory*, ed. Thomas L. Haskell. Bloomington: Indiana University Press.

Henderson, K. 1991. Flexible sketches and inflexible data bases: Visual communication, conscription devices, and boundary objects in design engineering." *Science, Technology, and Human Values* 16:448–473.

Henderson, R., A. Jaffe, and M. Trajtenberg. 1998. Universities as a source of commercial technology: A detailed analysis of university patenting, 1965–1988. *The Review of Economics and Statistics* 80(1):119–127.

Heydebrand, W. 1989. New organizational forms. *Work and Occupation* 16(3): 323–357.

Hicks, D. M., and J. S. Katz. 1996. Where is science going? *Science, Technology, & Human Value* 21(4):379–406.

Hughes, E. C. 1963. The professions. *Daedalus* (Fall) 92:655–668.

Hughes, T. P. 1989. The evolution of large technical systems. In *The social construction of technological systems*, ed. W. E. Bijker, T. P. Hughes, and T. J. Pinch, 51–82. Cambridge: MIT Press, Cambridge.

———. 1992. The dynamics of technical change: Salients, critical problems, and industrial revolutions. In *Technology and enterprise in a historical perspective*, eds. G. Dosi, R. Gianetti, P. A. Toninelli, 97–118. Oxford: Clarendon Press.

Institute of Medicine. 1985. *Personnel needs and training for biomedical and behavioral research*. Washington, DC: National Academy Press.

Jacoby, S., and P. Gonzales. 1991. The constitution of expert-novice in scientific discourse. *Issues in Applied Linguistics* 2(2):149–181.

Jaffe, A. 1989. Real effects of academic research. *American Economic Review* 79: 957–970.

Jaffe, A. B., M. Trajtenberg, and R. Henderson. 1993. Geographic localization of knowledge spillovers as evidenced by patent citations. *Quarterly Journal of Economics* 7(5):577–598.

Jasanoff, S., G. E. Markle, J. C. Petersen and T. Pinch, eds. 1995. *Handbook of science and technology studies*. Thousand Oaks, CA: Sage.

Jenkins, R. 1998. BT in the hot seat. *Seedling* 15:13–21.

Kaghan, W. N., and G. B. Barnett. 1997. The desktop model of innovation in digital media. In *Universities and the global knowledge economy: A triple helix of university-industry-government relations*, eds. H. Etzkowitz and L. Leydesdorff, 71–81. Leicester, UK: Pinter/Cassell/Leiscester University Press.

Kenney, M. 1986. *Biotechnology: The university-industrial complex*. New Haven: Yale University Press.

———. 1987. The ethical dilemmas of university-industry collaborations. Journal of Business Ethics 6:127–135.

Kennedy, D. 1997. *Academic duty*. Cambridge, MA: Harvard University Press.

Kerr, C. 1963. *The uses of the university.* New York: Harper and Row.

———. 1982. The uses of the university two decades after: Postscript 1982. *Change* 14:23–31.

Kevles, D. 1987. *The physicists: The history of a scientific community in modern America.* Cambridge, MA: Harvard University Press.

Keyworth, G. A., II. 1983. Federal R&D: Not an entitlement. *Science* 219:801.

Kleiner, K. 1995. Fear and loathing at the Smithsonian. *New Scientist* 8:42.

Kleinman, D. L. 1995. *Politics on the endless frontier: Postwar research policy in the United States.* Durham, NC: Duke University Press.

———. 1998. Untangling context: Understanding a university laboratory in the commercial world. *Science, Technology, & Human Values* 23(3):285–314.

Kleinman, D. L., and J. Kloppenburg. 1988. Biotechnology and university-industry relations: policy issues in research and the ownership of intellectual property at a land grant university. *Policy Studies Journal* 17:83–96.

Kloppenburg, J. 1988. *First the seed: The political economy of plant biotechnology, 1492–2000.* New York: Cambridge University Press.

Knorr-Cetina, K. 1999. *Epistemic cultures: How the sciences make knowledge.* Cambridge: Harvard University Press.

———. 1981. *The manufacture of knowledge: An essay on the constructivist and contextual nature of science,* Pergamon Press, Oxford.

Kogut, B. M., ed. 1993. *Country competitiveness: Technology and the organizing of work.* New York: Oxford University Press.

Kohler, R. 1990. *Partners in science: Foundations and natural scientists, 1900–1945.* Chicago: University of Chicago Press.

Kohn, G. 1987. Agriculture, pesticides, and the American chemical industry. In *Silent spring revisited,* ed. G. Marco, R. Hollingsworth, and W. Durham, 159–174. Washington, DC: American Chemical Society.

Krimsky, S., J. Ennis, and R. Weissman. 1991. Academic-corporate ties in biotechnology: A quantitative study. *Science, Technology, and Human Values* 16:275–288.

Kruytbosch, K., and S. Messinger. 1968. Unequal peers: The situation of researchers at Berkeley. *American Behavioral Scientist* 11:33–43.

Kunda, G. 1992. *Engineering culture: Control and commitment in a high-tech corporation.* Philadelphia, PA: Temple University Press.

Kutter, R. 1983. The declining middle. *Atlantic Monthly* (July):60–71.

La Follette, M. C., ed. 1982. *Quality in science.* Cambridge, MA: MIT Press.

Lanjouw, J. O., A. Pakes, and J. Putnam. 1996. *How to count patents and value intellectual property: Uses of patent renewal and application data.* National Bureau of Economic Research Working Paper #5741.

Latour, B. 1987. *Science in action: How to follow scientists and engineers through society.* Cambridge, MA: Harvard University Press.

———. 1988 *The pasteurization of France* . Cambridge, MA: Harvard University Press.

———. 1993. *We have never been modern*. Cambridge, MA.: Harvard University Press.

———. 1996. *Aramis, or the love of technology*. Cambridge, MA: Harvard University Press.

Latour, B., and S. Woolgar.1979/1986. *Laboratory life: The social construction of scientific facts*. Beverly Hills, CA: Sage.

Law, J. 1994. *Organizing modernity*. Oxford: Blackwell.

Law, J., and M. Callon. 1988. Engineering and sociology in a military aircraft project: A network analysis of technological change. *Social Problems* 35:284–297.

Law, J. 1989. On the construction of sociotechnical networks: Content and context revisited. *Knowledge and society: Studies in the sociology of science past and present* 8:57–83.

———. 1992. The life and death of an aircraft: A network analysis of technical change. In *Shaping technology/Building society*, ed. W. E. Bijker and J. Law, 21–52. MIT Press, Cambridge.

Lee, Y. S. 1996. Technology transfer and the research university: A search for the boundaries of university-industry collaboration. *Research Policy* 25:843–863.

Leslie, S. 1993. *The Cold War and American science: The military-industrial-academic complex at MIT and Stanford*. New York: Columbia University Press.

Leslie, S. W. 1987. Playing the education game to win: The military and interdisciplinary research at Stanford. *Historical Studies in the Physical and Biological Sciences* 18:55–88.

Levin, S. G., and P. E. Stephan. 1991. Research productivity over the life cycle. *American Economic Review* 81:114–132.

———. 1998. Gender differences in rewards to publishing in academe: Science in the 1970s. *Sex Roles* 38(11/12):1049–1064.

Levinson, R. M. 1989. The faculty and institutional isomorphism. *Academe* 75:27–27.

Lindblom, C. E. 1980. The privileged position of business in policy making. In *The policy making process*, 71–82. Englewood Cliffs, NJ: Prentice-Hall.

Lockeretz, W., and M. D. Anderson. 1993. *Agricultural research alternatives*. Lincoln: University of Nebraska Press.

Louis, K., and M. Anderson. 1998. The changing context of science and university-industry relations. In *Capitalizing knowledge: New intersections of industry and academia*, ed. H. Etzkowitz, A. Webster, and P. Healey, 73–94. Albany: SUNY Press.

Louis, K. S., D. Blumenthal, M. Gluck, and M. Stoto. 1989. Entrepreneurs in academe: An exploration of behaviors among life scientists. *Administrative Science Quarterly* 34:110–131.

Low, G. M. 1980. Keynote address to the National Academy of Engineering Symposium on Academe-Industry-Government Interaction in Engineering Education. Washington, DC, (30 October).

Macaulay, S. 1963. Non-contractual relations in business: A preliminary study. *American Sociological Review* 28:55–67.

Macilwain, C. 1995a. Now chemists hit at Smithsonian's 'antiscience' exhibit. *Nature* (April):752.

———. 1995b. Smithsonian heeds physicists' complaints. *Nature* 16:307.

MacKenzie, D. A. 1996. *Knowing machines: Essays on technological change.* Cambridge, MA: The MIT Press.

Mackenzie, M., P. Keating, and A. Cambrosio. 1990 Patents and free scientific information in biotechnology: Making monoclonal antibodies proprietary. *Science, Technology, and Human Values* 15:65–83.

Mansfield, E. 1991. Academic research and industrial innovation *Research Policy* 20:1–12.

———. 1995 Academic research underlying industrial innovations: Sources, characteristics, and funding. *The Review of Economics and Statistics* 54:55–65.

March, J. G. 1988. *Decisions and organizations.* Oxford: Basil Blackwell.

Marshall, E. 1997 Secretiveness found widespread in the life sciences. *Science* 276 (April 25):525.

Matkin, G. W. 1990. *Technology transfer and the university.* New York: MacMillan.

Merton, R. 1942/1973. The normative structure of science. In *The sociology of science: Theoretical and empirical investigations*, ed. R. Merton, 267–278. Chicago: University of Chicago Press.

Merton, R. K. 1957. Priorities in scientific discovery: A chapter in the sociology of science. *American Sociological Review* 22(6):635–659.

———. 1968. The Matthew Effect in science. *Science* 159(3810):56–63.

———. 1973. *The sociology of science: Theoretical and empirical Investigations.* Chicago: University of Chicago Press.

———. 1976. *Sociological ambivalence and other essays.* New York: The Free Press.

———. 1988 The Matthew Effect in science, II: Cumulative advantage and the symbolism of intellectual property. *Isis* 79(299):606–623.

Meyer, J. and B. Rowan. 1977. Institutionalized organizations: Formal structure as myth and ceremony. *American Journal of Sociology* 83:340–363.

———. 1978. The structure of educational organizations. In *Environments and organizations*, ed. M. W. Meyer, 78–109. San Francisco: Jossey-Bass.

Miller, D. L. 1989. *Lewis Mumford: A life.* New York: Weidenfeld and Nicolson.

Mishkin, B. 1995. Urgently needed: Policies on access to data by erstwhile collaborators. *Science* 270 (November 10):927–928.

Mitroff, I. I. 1974a. Norms and counter-norms in a select group of the Apollo Moon scientists: A case study of the ambivalence of scientists. *American Sociological Review* 39:579–595.

———. 1974b. *The subjective side of science.* Amsterdam, The Netherlands: Elsevier.

Morrill, C. 2000. Institutional change and interstitial emergence: The growth of alternative dispute resolution in American Law, 1965–1995. (Forthcoming). In *Bending the bars of the iron cage: Institutional dynamics and processes,* ed. W. W. Powell and D. L. Jones. Chicago: University of Chicago Press.

Mowery, D. C., and N. Rosenberg. 1989. *Technology and the pursuit of economic growth.* Cambridge, UK: Cambridge University Press.

———. 1993. The U.S. national innovation system. In *National innovation systems: A comparative analysis,* ed. R. R. Nelson, 29–75. Oxford: Oxford University Press.

Mulkay, M. 1979. *Science and the sociology of knowledge.* London; Boston: G. Allen & Unwin.

———. 1980. Interpretation and the use of rules: The case of the norms of science. *Transactions of the New York Academy of Sciences:*111–125.

Myers, G. 1995. From discovery to invention: The writing and rewriting of two patents. *Social Studies of Science* 25:57–105.

Narin, F. 1994. Patent bibliometrics. *Scientometrics* 30(1):147–155.

Narin, F., and A. Brietzman. 1995. Inventive productivity. *Research Policy* 24:507–519.

Narin, F., K. S. Hamilton, and D. Olivastro. 1997. The increasing linkage between U.S. technology and public science. *Research Policy* 26:317–330.

Narin, F., and D. Olivastro. 1992. Status report: Linkage between technology and science. *Research Policy* 21:237–249.

Narin, F., and D. Olivastro. 1998. Linkage between patents and papers: An interim EPO/US comparison. *Scientometrics* 41(1–2):51–59.

National Research Council. 1981. *Postdoctoral appointments and disappointments.* Washington, DC: National Academy Press.

———. 1982. *An assessment of research-doctorate programs in the United States: Biological sciences.* Washington, DC: National Academy Press.

———. 1998. *Trends in the early careers of life scientists.* Washington, DC: National Academy Press.

National Science Board. 1987. *Science and engineering indicators 1987.* Washington, DC: U.S. Government Printing Office.

———. 1989. *Science and engineering indicators 1989.* Washington, DC: U.S. Government Printing Office.

———. 1991. *Science and engineering indicators 1991.* Washington, DC: U.S. Government Printing Office.

———. 1996. *Science and engineering indicators—1996.* Washington, DC: U.S. Government Printing Office.

National Science Foundation, Division of Policy Research and Analysis. 1988. *Distribution of U.S. academic research activity: An economic view.* Washington, DC: mimeo (September).

Nelkin, D., and R. Nelson. 1987. Commentary: University-industry alliances. *Science, Technology, and Human Values* 12:65–74.

Nelson, R. R. 1977. *The moon and the ghetto.* New York: Norton.

———. 1989. What is private and what is public about technology. *Science, Technology, and Human Values* 14(3):229–241.

———. 1990. Capitalism as an engine of progress. *Research Policy* 19:193–214.

———. 1995. Why should managers be thinking about technology policy? *Strategic Management Journal* 16:581–588.

———. 1996. *The sources of economic growth.* Cambridge, MA: Harvard University Press.

Nelson, R. R., ed. 1993. *National innovation systems: A comparative analysis.* Oxford: Oxford University Press.

Nicholson, H. 1977. Autonomy and accountability of basic research. *Minerva* XV:32–61.

Noble, D. F. 1977. *America by design: Science, technology, and the rise of corporate capitalism.* Oxford: Oxford University Press.

———. 1986. *Forces of production: A social history of industrial automation.* New York: Oxford University Press.

Nohria, N., and J. Eccles, eds. 1992. *Networks and organizations: Structure, form and action.* Boston: Harvard Business School Press.

North, D. C. 1981. *Structure and change in economic history.* New York: Norton.

———. 1990. *Institutions, institutional change, and economic performance.* Cambridge: Cambridge University Press.

Office of Technology Assessment. 1988. *Educating scientists and engineers: Grade school to grad school.* Washington, DC: U.S. Government Printing Office.

Orlikowski, W. J. 1996. Improvising organizational transformation over time: A situated change perspective. *Information Systems Research* 7:63–92.

Osburn, R., J. Milner, E. Oplinger, R. Smith, and J. Handelsman. 1995. Effect of *Bacillus cereus* UW85 on the yield of soybean at two field sites in Wisconsin. *Plant Disease* 79:551–556.

Ostry, S., and R. R. Nelson. 1995. *Techno-nationalism and techno-globalism: Conflict and cooperation.* Washington, DC: Brookings Institution.

Owen-Smith, J. *New arenas for university competition: Accumulative advantage in academic patenting* [this volume].

Packer, K., and A. Webster. 1996. Patenting culture in science: Reinventing the scientific wheel of credibility. *Science, Technology, and Human Values.* 21(4):427–453.

Pakes, A., and Z. Griliches. 1980. Patents and R&D at the firm level: A first look. *Economics Letters.* 5:377–381.

Pearson, R. 1987. Researchers living on the fringe. *Nature* 327:86.

Pelz, D., and F. M. Andrews. 1976. *Scientists in organizations: Productive climates for research and development* (rev. ed.). Ann Arbor, MI: Institute for Social Research, University of Michigan.

Perrolle, J. A. 1986. Intellectual assembly lines: The rationalization of managerial, professional, and technical work. *Computers in the Social Sciences* 2:111–121.

Peters, L., and H. Fusfeld. 1982. *Selected studies in university-industry research relationships.* Washington, DC: National Science Foundation.

Pfeffer, J., and G. Salancik. 1978. *The external control of organizations.* New York: Harper and Row.

Phillips, D. I., and B. P. S. Shen, eds. 1982. *Research in the age of the steady-state university.* Boulder, CO: Westview.

Pisano, G. P. 1994. Knowledge, integration, and the locus of learning: An empirical analysis of process development. *Strategic Management Journal* 15:85–100.

Podolny, J., T. E. Stuart, and M. T. Hannan. 1996. Networks, knowledge, and niches: Competition in the worldwide semi-conductor industry. *American Journal of Sociology* 102(3):659–689.

Polanyi, M. 1951. *The logic of liberty: Reflections and rejoinders.* Chicago: University of Chicago Press.

———. 1962. The republic of science. *Minerva* 1:54–73.

Porter, M. E. 1985. *Competitive strategy: Creating and sustaining superior performance.* New York: Free Press.

———. 1990. *The competitive advantage of nations.* New York: Free Press.

Powell, W., K. Koput and L. Smith-Doerr. 1996. Interorganizational collaboration and the locus of Innovation: Networks of learning in biotechnology. *Administrative Science Quarterly* 41:116–145.

Powell, W. W., and P. J. DiMaggio, eds. 1991. *The new institutionalism in organizational analysis.* Chicago: University of Chicago Press.

Powell, W. W., and J. Owen-Smith. 1998. Universities and the market for intellectual property in the life sciences. *Journal of Policy Analysis and Management* 17(2): 253–277.

Powers, D. R. 1988. *Higher education in partnership with industry.* San Francisco: Jossey-Bass.

Price, D. 1984. The science/technology relationship, the craft of experimental science, and policy for the improvement of high technology innovation. *Research Policy* 13:3–20.

Rabinow, P. 1996. *Making PCR: A story of biotechnology.* Chicago: University of Chicago Press.

Ravetz, J. 1971. *Scientific knowledge and its social problems.* New York: Oxford.

Remington, J. A. 1988. Beyond big science in America: The binding of inquiry. *Social Studies of Science* 18:45–72.

Rhoades, G. 1998. *Managed professionals: Unionized faculty and restructuring academic labor.* Albany: State University of New York Press.

Rhoades, G., and S. Slaughter. 1991a. Professors, administrators, and patents: The negotiation of technology transfer. *Sociology of Education* 64(2):65–77.

———. 1991b. The public interest and professional labor: Research universities. In *Culture and ideology in higher education,* ed. W. Tierney, 187–212. New York: Praeger.

———. 1997. Academic capitalism, managed professionals, and supply-side higher education. *Social Text,* 15:9–38.

Rip, A. 1986. Mobilizing resources through texts. In *Mapping the dynamics of science and technology,* ed. M. Callon, J. Law, and A. Rip, 84–99. London: Macmillan Press.

Rip, A., and A. J. Nederhof. 1986. Between dirigisme and laissez-faire: Effects of implementing the science policy priority for biotechnology in The Netherlands. *Research Policy* 15:253–268.

Roberts, S., ed. 1984. *Academic research in the United Kingdom: Its organization and prospects.* London: Taylor Graham..

Rosenberg, N. 1990. Why do firms do basic research (with their wwn money)? *Research Policy* 19:165–174.

Rosenberg, N., and R. R. Nelson. 1994. American universities and technical advance in industry. *Research Policy* 23:323–348.

Rosenburg, S. A. 1996. Secrecy in medical research. *New England Journal of Medicine,* 334(6):392–394.

Rosengrant, S., and D. Lampe. 1992. *Route 128: Lessons from Boston's high tech community.* New York: Basic Books.

Rumelt, R. P. 1974. *Strategy, structure and economic performance.* Boston, MA: Harvard Business School Press.

Salaman, G. 1983. Managing the frontier of control. In *Social class and the division of labor,* ed. A. Giddens and G. MacKenzie, 46–62. Cambridge: Cambridge University Press.

Samuelson, P. 1987. Innovation and competition: Conflicts over intellectual property rights in new technologies. *Science, Technology, and Human Values* 12:6–21.

Sawyer, R. 1996. *To make a spotless orange: Biological control in California.* Ames, IA: Iowa State University Press.

Saxenian, A. 1994. *Regional advantage.* Cambridge, MA: Harvard University Press.

Schmookler, J. 1966. *Invention and economic growth.* Cambrige, MA: Harvard University Press.

Schultz, J. 1996. Interactions between university and industry. In *Biotechnology,* ed. F. B. Rudolph and L. W. McIntire, 131–146. Washington, DC: Joseph Henry Press.

Schulman, S. 1999. *Owning the future.* New York: Houghton Mifflin.

Scott, W. R. 1995. *Institutions and organizations*. Newbury Park, CA: Sage.

Seashore, K. L., D. Blumenthal, M. Gluck, and M. Stoto. 1989. Entrepreneurs in academe: An exploration of behavior among life scientists. *Administrative Science Quarterly* 34:110–131.

Shapin, S. 1988. The house of experiment in 17th century England. *ISIS* 79(298): 373–404.

Shapley, D., and R. Roy. 1985. *Lost at the frontier: U.S. science and technology policy adrift*. Philadelphia: ISI Press.

Shenk, D. 1999. Money+science = ethics problems on campus. *The Nation* (March 22):11–18.

Shils, E. 1983. *The academic ethic*. Chicago: University of Chicago Press.

Slaughter, S. 1988. Academic freedom and the state. *Journal of Higher Education* 59:241–262.

———. 1990. *The higher learning and high technology: Dynamics of higher education policy formation*. Albany, NY: SUNY Press.

———. 1993. Beyond basic science: Research university presidents' narratives of science policy. *Science, Technology, and Human Values* 18(3):278–302.

———, and G. Rhoades. 1990. Renorming the social relations of academic science: Technology transfer. *Educational Policy* 4:341–361.

———. 1991. Professors, administrators, and patents: The negotiation of technology transfer. *Sociology of Education* 64(2):65–78.

———. 1993. Changes in intellectual property statutes and policies at a public university: Revising the terms of professional labor. *Higher Education* 26:287–312.

———. 1996. The emergence of a competitiveness research and development policy coalition and the commercialization of academic science and technology. *Science, Technology, and Human Values* 21(3):303–339.

Slaughter, S., and L. L. Leslie. 1997. *Academic capitalism: Politics, policies and the entrepreneurial university*. Baltimore: John Hopkins University Press.

Smiley, J. 1995. *Moo*. New York: Knopf.

Smith, B.L.R. 1985. Graduate Education in the United States. In *The state of graduate education*, ed. B.L.R. Smith, 1–30. Washington, DC: The Brookings Institution.

———. 1990. *American science policy since World War II*. Washington, DC: The Brookings Institution.

Smith-Doerr, L., J. Owen-Smith, K. W. Koput, and W. W. Powell. 1999. Networks and knowledge production: Collaboration and patenting in biotechnology. In *Corporate Social Capital*, ed. R.Th.A.J. Leenders and S. M. Gabbay, 390–408. Boston: Kluwer.

Sokolowski, R. 1991. The fiduciary relationship and the nature of professions. In *Ethics, trust, and the professions*, ed. E. D. Pellegrino, R. M. Veatch, and J. P. Langan, 23–43. Washington, DC: Georgetown University Press.

Star, S. L. 1989. The structure of ill-structured solutions: Boundary objects and heterogenous distributed problem solving. In *Distributed artificial intelligence*, Vol. 2, ed. M. Huhns and L. Gasser, 37–54. San Mateo: Morgan Kaufman.

Steelman, J. R. 1947. *Science and public policy: A report to the President.* Washington DC: U.S. Government Printing Office.

Stehr, N. 1994. *Knowledge societies.* London: Sage Publications.

Stephan, P. E. 1996. The economics of science. *Journal of Economic Literature.* 34:1199–1235.

Stern, B. J. 1954. Freedom of research in American science. *Science and Society* XVIII:97–122.

Stinchcombe, A. L. 1990. *Information and organizations.* Berkeley: University of California Press.

Strathern, M. 1999. What is intellectual property after? In *Actor network theory and after*, ed. J. Law and J. Hassard, 156–180. Oxford, UK: Blackwell Publishers/The Sociological Review.

Storer, N. W. 1966. *The social system of science.* New York: Holt, Rinehart, and Winston.

Strauss, A. L. 1978. Negotiations: Varieties, contexts, processes, and social order. San Francisco: Jossey-Bass.

———. 1988. The articulation of project work: An organizational process. *Sociological Quarterly* 29:163–178.

Teece, D. J. 1986. Profiting from technological innovation. *Research Policy* 15: 286–305.

Teitelman, R. 1989. *Gene dreams: Wall Street, academia, and the rise of biotechnology.* New York : Basic Books.

Teich, A. 1982. Research centers and non-faculty researchers: A new academic role. In *Research in the age of the steady-state university,* ed. D. I. Phillips and B.P.S. Shen, 91-108. Boulder, CO: Westview.

Thompson, J. 1967. *Organizations in action.* New York: McGraw-Hill.

Tolbert, P. 1985. Institutional environments and resource dependence: Sources of administrative structure in institutions of higher education. *Administrative Science Quarterly*, 30:1–13.

Trajtenberg, M. 1990. A penny for your quotes: Patent citations and the value of Information. *Bell Journal of Economic.* 21:172–187.

Traweek, S. 1989. *Beamtimes and lifetimes: The world of high energy physicists.* Cambridge: Harvard University Press.

U.S. General Accounting Office. 1994. *Peer review: Reforms needed to ensure fairness in federal agency grant selection* (No. PEMD-94-1). Washington, DC: U.S. Congress.

U.S. Congress. 1991. Senate, Committee on Commerce, Science, and Transportation, Subcommittee on Science, Technology, and Space, NASA's plan to restructure the space station freedom: Hearing before the subcommittee on Commerce, Science, and Transportation. U.S. Senate, One Hundred Second Congress, first session, *Cong. Rec.* April 16. Washington: DC: G.P.O.

U.S. Congress, House of Representatives. 1993. Academic Earmarks—Part III. Hearings before the Committee on Science, Space, and Technology, September 21, 22 and October 6.

USPTO (United States Patent and Trademark Office). 1997. *U.S. colleges and universities—Utility patent grants 1969–1997.* Washington, DC: U.S. Department of Commerce.

Utterback, J. M. 1994. *Managing the dynamics of innovation.* Boston: Harvard Business School Press.

Uzzi, B. 1996. The sources and consequences of embeddedness for the economic performance of organizations: The network effect. *American Sociological Review* 61:674–698.

Vavakova, B. 1996. The new social contract: Governments, universities, and society, presented at The Triple Helix: University-Industry-Government Relations. Amsterdam, the Netherlands, January 4–6, 1996.

Veblen, T. 1918 (Rpt 1954). *The higher learning in America, A memorandum on the conduct of universities by businessmen.* Stanford, CA: Academic Reprints.

Weber, M. 1918 (Rpt. 1946). Science as a Vocation. In *From Max Weber: Essays in Sociology,* ed. H. Gerth and C. W. Mills, 129–156. New York: Oxford University Press.

Webster, A. 1994. University-corporate ties and the construction of research agendas. *Sociology* 28:123–142.

Webster, A., and K. Packer. 1997. Patents in public sector research: When worlds collide. In *Universities and the global knowledge economy: A triple helix of university-industry-government relations,* eds. H. Etzkowitz and L. Leydesdorff, 47–59. Leicester, UK: Pinter/Cassell/Leiscester University Press.

Weick, K. E. 1976. Educational organizations as loosely coupled systems. *Administrative Science Quarterly* 21:1–19.

———. 1979. *The social psychology of organizing.* Reading, MA: Addison-Wesley.

———. 1990. Technology as equivoque: Sensemaking in new technologies. In *Technology and organizations,* ed. P. S. Goodman and L. S. Sproull and Associates, 1–44. San Francisco: Jossey-Bass Publishers.

———. 1998. Improvisation as a mindset for organizational analysis. *Organization Science* 9:543–555.

Weidman, J. C. 1989. Undergraduate socialization: A conceptual approach. In *Higher education: Handbook of theory and research,* ed. J. C. Smart, 289–322. New York: Agathon Press.

Willis, P. 1977. *Learning to labor.* Aldershot: Gower.

Weiner, C. 1986. Universities, professors, and patents: A continuing controversy. *Technology Review* 89 (February/March).

Weiner, C. 1987. Patenting and academic research: Historical case studies. *Science, Technology, and Human Values* 12:50–62.

Williamson, O. E. 1985. *The economic institutions of capitalism*. New York: The Free Press.

Winner, L. 1977. *Autonomous technology: Technics-out-of-control as a theme in political thought*. Cambridge, MA: The MIT Press.

———. 1986. Brandy, cigars, and human values. In *The whale and the reactor*, 155–163. Chicago: University of Chicago Press.

Wright, E. O., and J. Singlemann. 1982. Proletarianization in the changing American class structure. *American Journal of Sociology* 88 (Supplement):5176–5209.

Wyngaarden, J. B. 1984. Nurturing the scientific enterprise. *Science* 223:361–364.

Zachary, G. P. 1997. *Endless frontier: Vannevar Bush, engineer of the American century*. New York: The Free Press.

Zenzen, M., and S. Restivo. 1982. The mysterious morphology of immiscible liquids: A study of scientific practice. *Social Science Information* 21:447–473.

Zerubavel, E. 1991. *The fine line: Making distinctions in everyday life*. New York: Free Press.

Zuboff, S. 1988. *In the age of the smart machine*. New York: Basic.

Zucker, L. 1989. Comment on Carroll-Hannan. *American Sociological Review* 54:542–545.

Zucker, L. G., M. R. Darby, and M. B. Brewer. 1994. *Intellectual capital and the birth of U.S. biotechnology enterprises*. National Bureau of Economic Research, Working Paper 4653.

Zuckerman, H. 1977. *Scientific elite: Nobel laureates in the United States*. New York: Free Press.

Zussman, R. 1985. *Mechanics of the middle class: Work and politics among American engineers*. Berkeley: University of California Press.

# Contributors

JENNIFER L. CROISSANT is assistant professor in the Program on Culture, Science, Technology, and Society at the University of Arizona. Current interests include research projects on technology and organizational change, cultural constructions of the real and the natural, and work on scientific instrumentation, particularly in archaeology.

EDWARD J. HACKETT is professor of sociology at Arizona State University. His recent work concerns the dynamics of research groups in science and engineering, peer review, and misconduct in science. From 1996 to 1998 he directed the Science and Technology Studies Program at the National Science Foundation.

DEBORAH G. JOHNSON is professor and director of the Program on Philosophy, Science, and Technology in the School of Public Policy at Georgia Institute of Technology in Atlanta, Georgia. She is currently working on the third edition of her popular textbook *Computer Ethics*. Johnson's research interests include computer and engineering ethics, professional ethics, and values and policy. Her most recent publications focus on democratic values and the Internet, and forbidden knowledge in modern science.

WILLIAM N. KAGHAN is a consultant/contractor in technology management and technical communication in affiliation with Sakson & Taylor, Inc in Seattle, Washington. He worked for eight years as a research and development engineer at the Boeing Corporation. His dissertation centered on a participant observer field study at the University of Washington Office of Technology Transfer. He continues to research the everyday work of members of occupational groups such as technology transfer managers and technical communications specialists who play an important, but often invisible, role in the process of technological innovation.

DANIEL LEE KLEINMAN is associate professor of sociology in the School of History, Technology, and Society at the Georgia Institute of Technology. He is author of *Politics on the Endless Frontier: Postwar Research Policy in the United*

*States* (Duke, 1995) and the editor of *Science, Technology, and Democracy* (SUNY, forthcoming). His current research uses ethnographic data to explore the ways in which the commercial world shapes the daily practices of university biologists.

W. PATRICK MCCRAY received his original training as a physical scientist and later did his doctoral work in the history of technology. He also developed an interest in the application of ethnography in researching contemporary science and technology and is now a research fellow at The Center for History of Recent Science at The George Washington University. His current work is on the interaction of "science" and "technology" in the design, development, and operation of contemporary astronomical instrumentation, including the Gemini 8-Meter Telescopes and the Next Generation Space Telescope.

JEFFREY L. NEWCOMER is currently an assistant professor of manufacturing engineering technology in the Engineering Technology Department at Western Washington University. He received his B.S. and M.S. in aeronautical engineering, a Ph.D. in mechanical engineering, and an M.S in science and technology studies from Rensselaer Polytechnic Institute.

JASON OWEN-SMITH is a doctoral candidate in sociology at the University of Arizona. His dissertation mixes quantitative and case study methods in an analysis of the causes and consequences of patenting by Research One universities. After graduation in August 2000 Owen-Smith plans to conduct postdoctoral research at Stanford University.

SAL RESTIVO is professor of sociology and science studies at Rensselaer Polytechnic Institute in Troy, New York. He is a founding member and past president of the Society for Social Studies of Science and has published widely on the sociology of science, mathematics, and mind.

BLAIR SCHNEIDER is currently the network administrator for Fastech, a software development company in Livonia, Michigan. The firm specializes in "Time and Attendance" software, producing electronic timesheets and computer-based management aids. Previously he worked as consultant in and around U.S. automobile manufacturers after graduate experience in science and technology studies at Rensselaer Polytechnic Institute.

LANGDON WINNER, professor of political science in the Department of Science and Technology Studies at Rensselaer Polytechnic Institute, is a political theorist who focuses upon social and political issues that surround modern technological change. He is author of *Autonomous Technology* and *The Whale and the Reactor: The Search for Limits in an Age of High Technology* and is currently doing research and writing a book about the politics of design in the contexts of engineering, architecture, and political theory. Past president of the Society for Philosophy and Technology, he was also a contributing editor at *Rolling Stone* in the late 1960s and early 1970s.

SUNY series in
Science, Technology, and Society
Sal Restivo and Jennifer Croissant, editors

List of Titles:

# Index